The Projecting Brain

Understanding how the brain projects conscious awareness to help us thrive in an uncertain world

James Vickers

Leyton Works
Hobart

The Projecting Brain

Published by Leyton Works

ISBN 978-1-7647982-1-1
ISBN (e-book) 978-1-7647982-0-4

Cover figure by Chris Rees

My thanks to all who have helped in the development of this book.

Cerebrum de cerebro cogitat

CONTENTS

Chapter 1 The miraculous pink blob

Projected into two worlds at once

I am in the garden – weeding, shovelling, and enjoying the audiobook of The Fellowship of the Ring *(BBC version), touchstone of proud nerds. I am in our garden, but as I listen to the words, I am also in Middle Earth (in the tunnels of Moria). My brain slips seamlessly between two parallel realities, neither requiring focused mental work. My mind is projected into sifting weeds and clumps of dirt, and also into the chilling claustrophobia of tunnels in the dark, following a wizard who seems to know where he is going....*

The same phenomenon works for us on multiple levels every day. Driving a car, for example– driving along, we experience the feel of tyres on the road, the melody of a song on the radio, the position of the car relative to the road and the external world, and a subtle prediction of what will happen if we follow our current path. We are projected into this environment: our brain immerses us into the external world, and this experience is orchestrated to a rhythm of time in a shifting place, a seamless existence created by the most marvellous biological machinery, which – when working well– doesn't feel like anything at all.

Who we are – physically – is a brain in a bone box, but the golden trick of the brain is to continuously process bountiful sensory and cognitive information to project us into an external environment. Our mental state is the projection into these spaces, whether they are 'real' (my garden) or imagined (Middle Earth). When the neurobiology of this projection breaks down, perhaps as a consequence of mental illness, brain injury, or the gradual degeneration of brain disease, our ability to provide an accurate projection into the external world can be impacted, impeding our interactions with other people and our surroundings. But when it works well, our ability to project gives us the capacity to live within other worlds, such as a book or movie, or even through alternative identities online. Our marvellous brains allow us to colonise both physical and imagined spaces. To thrive, we must nurture and preserve this extraordinary capacity. Protecting our brains from the impacts of technology and neurodegeneration will be key to a healthy future.

Most people likely don't think at all about their brain. Some may even screw up their face with a disgusted look when reminded that, in some sense, everything they think and perceive originates in the blob-like mass between their ears. If you are reading this book, there is a good chance you are among the minority who does think about their brain.

The majority may spend a large proportion of their day thinking about how they look, what their hands are doing, whether their stomach is rumbling, if their back is sore, whether their arthritis is playing up, or if their eyes feel tired –

along with a vast array of other feelings and sensations arising from the body. But the knowledge that these functions are performed by an organ of the body is not front and centre of daily awareness. We're continually reminded of the essential roles of our heart, lungs and gut, for example, but unless we have a headache, we may not even consider pink blob inside our cranium. Why is that?

We may from time to time consider our internal monologue – the constant chatter, often repeating, returning to particular themes in a circular fashion – but do we attempt to access or focus on the stuff that generates such internal thought? Not so much.

Your brain needs a lot of blood, consumes a lot of energy, and generates electricity – but can you feel it doing those things? Just a quick little exercise: put down this book, close your eyes and see if you can pick up any sensation or feeling of your brain actually working. Give that a minute or two. I'll leave the rest of the page blank so you are not distracted...

Welcome back. What you probably got by way of internal sensation was more of the internal monologue, perhaps some flitting through images or memories, but nothing directly linked to the electrical and chemical activity of billions of nerve cells, their multiple billions of connections, or their respective trillions of synaptic connections. You experience their function, most often seamlessly, but unlike the beating of your heart or the sensation of your lungs inflating and deflating, you can't feel them working.

There must be a reason for this lack of direct experience of the brain – driven by aeons of the genetic changes that underlie evolution. The human brain, in particular the outer layer known as the cerebral cortex, sits at the pinnacle of biological evolution. It must work the way it does due to the selection pressures that shaped its functions and operations – resulting in the marvellous organ that provides us with understanding of the world around us, the capacity to communicate with other humans and to have internal thoughts and feelings which only the individual has access to. The brain provides the substrate for our culture, for our capacity to create, interact with and dominate our environment. Overseeing this exquisite capacity, we have consciousness – the personal experience of the world and our internal mental states. We experience the conscious state but it is a slippery, water-between-the-fingers phenomenon that – when working well – is hard to define and describe, both in terms of what it is and what it feels like when it is working.

An age-old mystery

There is no satisfactory explanation for how consciousness works. Ever since the birth of human philosophy, our species has pondered the inner mental world. As the 17th-century philosopher Renee Descartes observed, 'Cogito ergo sum!' – I am thinking therefore I am something, a 'self'! In modern times, the philosophy of consciousness is a vibrant field that is concerned with matters such as the substance of self-awareness and internal mental worlds, and with questions such as whether we are fooling ourselves into thinking that consciousness exists at all, and even whether consciousness deserves to be considered as its own fundamental property such as 'time' and 'space'.

Descartes notes that he can follow his process of thinking, and that he is convinced he has an internal mental world, but he can't describe how this mental world feels as it works. In the absence of a biological perception of the brain working, could it be that our rich mental world isn't down to cells, the connections between them, electricity and chemicals,? Perhaps that there is something else– a soul or spirit, an extradimensional entity that underlies our awareness? Some philosophers argue that while the brain and the internal mental world are equally real, they are not the same 'substances' but must be related in some intimate way.

Neuroscience takes us down a different path: it promises that the brain will ultimately be so well understood, we will understand how it performs its various functions including conscious awareness. Neuroscience is a very exciting field of human endeavour, involving thousands of scientists across the world, and, collectively, we have done an excellent job of discovering a tremendous amount about various brain regions, circuits and systems. Most of this new knowledge is akin to understanding the plumbing, the circuit boards, the highways and byways of the 1 to 1.5 kg of fat and protein that comprises our nervous system within the skull. We still lack a full picture of the link between the biochemical, cellular, physiological and anatomical components and our internal mental world of emotions, decisions, personality and morality. That doesn't mean we won't get there eventually, but we certainly haven't done that yet.

When the brain goes wrong

Many of us remain unaware of the pink blob that manages our inner mental world, but its existence becomes apparent, often tragically, when our brains no longer work so well. An acquired form of brain injury, such as an ischemic stroke as a result of a blocked blood vessel, can significantly impair particular functions of the brain, causing what we might describe as either a focal impairment (for example, difficulty moving a particular limb) or a global impairment (such as persistent feelings of fatigue). Similarly, a traumatic brain injury involving concussion, a loss of consciousness, can lead to impairments in logical thinking and working memory that persist over time. For the patient, describing this kind of deficit precisely can be difficult: “My brain doesn’t work so well anymore”. Such an injury may also cause an alteration in the wiring of the brain, setting off a heightened risk for abnormal electrical storms across neuronal networks, an epileptic seizure – which can also temporarily turn off usual awareness of surroundings.

Another form of brain dysfunction include mental health conditions such as depression, which flattens out the joys of life, often leading to despondency, negative interpretations of external events, and reduced social engagement.

Then there are ageing-related diseases such as Alzheimer’s disease that involve a gradual disconnection between regions of the cerebral cortex, contributing to an evolving and relentless deterioration in many of our higher cognitive functions, as well as our ability to connect with the external world. In a perverse twist, Alzheimer’s disease also often robs the individual of the powerful mental capacity of insight, such that the experience of deficit may be as bewildering to the person with this form of dementia as it is to the family members and friends around them.

Many of us will experience a serious mental health condition in our lifetime, and there will be many of our friends and family who also face this challenge. As average lifespan increases, more people will be exposed to the risk of neurodegenerative diseases such as dementia. Around 40-50% of young women born in Australia, for example, can expect to develop dementia in their lifetime. Another 40-50% will likely be in a caring role for someone with dementia.

Have we understood and mastered the major diseases and conditions that affect the *human* brain? Alas no, but why not? We now know a tremendous amount about the structure and function of the nervous system – particularly the 'simpler' neural machinery of nonhuman animals such as mice and invertebrates. We also have insights into the genetics and pathology of many diseases that affect human brains, again mainly through the use of animal models. We have also 'cured' many of these diseases and conditions in such experimental models. Lucky mice, rats and flies. Perhaps our ability to make a major leap forward in preventing and treating human brain diseases and conditions is being hampered by our failure to see the bigger picture of how the brain works: we know more and more about the components rather than the 'whole'.

Understanding this bigger picture may also provide insights into where we are headed as a species. Human beings are very creative organisms, developing culture and a variety of new ways to interact with the environment. These have included technological leaps such as in transportation, communication, information sharing and networking. At many points in our daily lives, our biology and our technological innovation come together, and our modern age is at the edge of an escalation of this integration between the brain and the machine worlds. We have a myriad of fascinating modern tools such as AI and brain-machine interfaces, but do we really understand how we work so well with them, and what may be the consequences of technological development into the future?

In this book, we will look at some of the major ideas about how our highly evolved brains work. I will also attempt to explain the 'why' and 'how' of a brain, particularly at the conscious level. This will rely on considering the sequence of gradual improvements in the sophistication of brain function in the course of evolution that culminated in the extraordinary capacity of our nervous system to integrate into the external environment – a capacity that may also explain why we have such little access to the brain's inner workings. Central to this is a new idea about how consciousness functions as a way for the brain to colonise and project into the world around it.

We will then explore how an understanding of the 'why' and 'how' of brain function may provide a new perspective on how humans relate to the environment and technology. Our brains are so adept at working through our machines and tools, to the extent that the human mind and technology may one day be fully integrated. A future to both thrill and terrify!

To end on something a bit less terrifying, we will finally consider how a wonderful human cultural invention – education – plays such an important role in shaping our lives and in realising the potential of what a brain can do. Education may also prove to be excellent medicine for what ails the brain.

The human brain is undoubtedly one of the most complex, mystifying structures in the known universe. Cracking its mysteries – not just its biological plumbing but the origin and purpose of consciousness – will be key to the future success of our species.

Chapter 2 A crash course in brain anatomy

If someone were to pluck your brain out and plonk it on the bench in front of you, the pinkish blob, crisscrossed with a spider web of blood vessels, would gradually flatten out and spread across the surface. The debrained you, of course, would no longer have the ability to be shocked or feel revulsion at the mess of blood, protein and fat spreading across the bench. Those assembled for the spectacle, however, might well be both fascinated and somewhat disappointed that this fragile substance, with the consistency of thick custard, was essentially the person they knew. They were used to your face, your body, your language, personality and mannerisms – but this is the first time they have met the real you. They might well wonder: "Is that all there is?".

Consider the neuron

If we were to scoop up part of that brain for analysis, it all becomes a lot more marvellous. Looking at the DNA, we would find thousands of genes responsible for developing a brain out of not much at all, conducting their business over many years to grow and sculpt a nervous system that is likely the most sophisticated thing in the universe. We would also find complex biochemical pathways that regulate the specific functions of nervous tissue, including the packaging, transport and release of a substantial range of chemical messengers, known as neurotransmitters, that then interact with a myriad of receptors.

If we were to take some pieces of brain and slice them very thinly, we could look at its constituent cells, including neurons, which transmit information around the brain, and several other types we'll come to shortly. Neurons are in many ways just like other cells of the body – they have a central nucleus containing genes; they have molecular machines in the cell body for producing and shaping proteins of various kinds; they have other biological machines for taking oxygen and nutrients supplied by blood cells and turning them into molecules that provide the fuel to run a range of metabolic processes; and they have an internal protein 'skeleton' of fine filaments for moving things around and giving the neuron its shape.

Perhaps the most unique feature of neurons that distinguishes them from other cells of the body is their shape. Neurons have multiple branch-like 'processes' – picture a deciduous tree in the middle of winter –typically that extend from the cell body. These include a number of branching structures called dendrites that are specialised to receive information from other nerve cells. They also have an axon, usually a single long branch that extends away from nerve cell body to either other nerve cells in the local vicinity, or over many millimetres or centimetres to interact with neurons at a distance. A combination of electrical and chemical action transmits information between two nerve cells via junctions known as synapses. The first cell releases neurotransmitters into the gap in the synapse and these bind to receptors in the membrane of the target neuron, affecting its relative excitability. When this reaches a threshold, an electrical signal known as an 'action potential' is propagated and passes down the axon to synapses at the end of this process which are stimulated to release their own neurotransmitters in turn. In this way, communication between neurons has both electrical (axon potential) and chemical (release of neurotransmitters) components.

There are a host of different neurotransmitters that bind to specific types of receptors. Some receptors are 'excitatory', pushing the neuron towards firing off an action potential, while others are 'inhibitory' and decrease the likelihood of an action potential. Other neurotransmitters may bind to receptors that have impacts on the internal workings of the neuron.

Very careful reconstructions of individual neurons have shown how substantially interconnected they are, with a single nerve cell connecting to dozens – sometimes even thousands – of other nerve cells. Each nerve cell also processes incoming stimulation in particular ways, depending on factors such as their complement of receptors for particular neurotransmitters and what part of the nerve cell is stimulated. This enables sophisticated processing of information at this cellular level before being passed on to the nerve cells to which they are connected.

Other types of brain cells

Neurons aren't the only cell type that comprise brains. Another major cell type are astrocytes, which can have lots of processes extending throughout their local environment and likely outnumber the neurons several-fold. Their processes can also enfold the blood vessels of the brain. Astrocytes have a variety of roles in the brain, from guiding patterns of neuronal growth in development, regulating blood flow, furnishing energy-containing molecules to neurons, and modulating neurotransmitters and synaptic transmission. Astrocytes may also play an important role in immune-related changes in the brain as well as in the defence of the nervous system following various forms of injury. In this latter regard, they can become 'reactive', proliferating and meshing together to wall off damaged areas of the brain, or defeating the spread of microorganisms.

Another important cell type in the brain are the microglia, that likewise have important roles in staving off infection and damage to the brain, as well as enabling disposal of microorganisms and broken-down cellular material. Microglia likely also have important roles in helping to 'sculpt' the processes and synaptic connections within the brain during its development as well in the adult brain.

Neuronal clumps and cortices

The brain has a symmetry, with patterns of organisation largely mirrored on both sides of the central axis. The largest structure known as the cerebrum, for example, is divided into two hemispheres. Nerve cells can also be organised into clumps throughout the brain, referred to as 'nuclei'. In addition there are sheets of cells that form an outer layer or 'cortex', such as the convoluted outer surface of the two hemispheres of the cerebrum, known as the cerebral cortex. The cerebellum, a smaller structure that sits above the brain stem, is also encased in a cortex. A substantial proportion of the brain is comprised of bundles of axons which make the 'white matter' of the brain, whereas the nerve cell bodies are largely located in the 'grey matter'. Axons can vary substantially in their diameter, which can influence how fast they can conduct an action potential – the wider the axon is, the faster the propagation of the action potential along the axon. Axons can be enveloped by a sheath of fatty membranes, known as myelin, that increases the speed of transmission of the action potential down the axon. Overall, the brain contains more fat (50-60% by weight) than any other constituent element, including proteins.

Looking closer at the 'architecture' of the cerebral cortex, it consists largely of columns of cells – including aligned bundles of dendrites and axons – organised in six distinct layers. It is thought that these columns represent the essential processing units of the cortex. Most of the neurons in these microcolumns are excitatory and can be relatively large, with axons that can extend to nearby neurons – referred to as collaterals – as well as longer projections extending many centimetres within the cortex or between the hemispheres. The remaining 20-30% of neurons in the cortex are inhibitory, and these vary both in their shape and how they physically interact with other neurons, with most of their connections being relatively short, extending to local neurons.

Which bits do what?

Research in experimental animals and in humans whose cortex has been damaged by traumatic injury or strokes suggests that brain functions are broadly associated with particular regions of the cerebral cortex. There are regions that can be described as sensory cortex, for example, where specific modalities of sensation, such as sight, hearing, touch and taste, are localised. There are regions that appear to have a role in combining and integrating different sensory modalities, known as association cortex. Also, there are regions that are linked with planning and executing movement: the premotor and motor cortex, respectively. Other areas of the cortex, particularly the frontal areas, appear more associated with characteristics such as personality and behaviour.

A particularly remarkable feature of the motor and sensory cortex is that their activity can be mapped to their roles in the body. For example, the primary motor cortex – which drives intentional movement – has a topographic representation to our muscles that extends over the surface of the hemispheres. Likewise, our somatosensory cortex, – relating to our sense of touch – is also mappable to our body regions including peripheral extremities. These maps have been described as 'homunculi', or little men, with a proportionate representation within the brain of particular parts of the body, such as relatively large areas for our hands representing sophisticated movement in the motor cortex, and sensitivity to touch in our somatosensory cortex.

For those of us who are neuroanatomically inclined, the other striking feature of the cerebral cortex is how similar one part of the cortex looks to other parts, with some subtle cellular variations. It is possible to delineate dozens of different regions of the cerebral cortex by the width of grey matter and differences in density and size of cells across its six layers, sometimes referred to as its 'cytoarchitecture'. This anatomic 'parcellation' of the cerebral cortex into subregions is remarkably well correlated with the broad localisation of several brain functions. Older studies have shown that functional deficits as a result of localised damage to these brain regions, for example as a result of strokes, traumatic brain injury and other forms of acquired damage, track broadly to these cortical regions, which can also be defined architecturally by how the cells and layers are organised. Additionally, there have been studies recording electrical activity in brain regions in awake individuals during brain surgery following various kinds of stimulation, as well as studies in which zapping regions with small amounts of current results in specific bodily sensations or kinds of movement. This has been supported in more recent decades by brain imaging studies that typically correlate variations in blood flow to relative neural activity when research subjects are stimulated in a variety of ways. Further understanding has come from comparative studies of brain structure and function across a range of other species.

Localising functions to particular brain regions in humans has been most successful in the context of cortical regions that are responsible for initiating motor activity (such as purposeful actions of our muscles), and for the processing of sensory information. The remarkable aspect of this is that, barring some variations in the size of cells and layers, the part of the cerebral cortex that is responsible for pointing your finger is almost indistinguishable in its makeup of cells from the cortical region that responds to a fly moving across your visual field, the experience of hearing a brass band, or the feel of a toddler touching your face. How can such similar clumps of cells give rise to such different functions and sensations?

Columns of cells in the cortex appear to act as some kind of core local processing unit. The processes of small groups of neurons, particularly pyramid-shaped or 'pyramidal' nerve cells, are oriented perpendicular to the outer surface of the cortex, aligning also with incoming bundles of axons. Are they really processing units or just a way of organising cells and neuronal processes? Neuroscience research has indicated that for cortical regions linked with visual processing they are often responsive to specific features of sensory information, such as the orientation of elements in the visual field.

There are also cortical regions that are more closely associated with 'primary' sensory functions: there are visual, auditory, gustatory (taste), olfactory (smell) and somatosensory (touch) cortices in each hemisphere of the brain. Remarkably, these individual sensory experiences 'feel' different to us, but the underlying cellular machinery looks very similar across these cortical regions, down to the columns of cells that comprise the grey matter. Only a skilled neuroanatomist could tell the difference between cortical regions linked to seemingly different perceptual experiences.

How about connectivity across larger distances? Does that tell us anything important about how the 'pink blob' performs the remarkable task of creating awareness? In a very broad sense, there are specific patterns of connection to and within the cerebral cortex. Firstly, it would seem to be really important that the two hemispheres are interconnected. Axons from pyramidal cells on one side of the brain extend across the hemispheres through bands of fibres known as the corpus callosum. In this regard, similar regions in each hemisphere are substantially interconnected through this corpus callosum. While it is uncommon, it is possible to be born and for the brain to develop without this substantial pathway of inter-hemisphere connectivity. This can be associated with motor and neuropsychological problems, but there are also individuals without a corpus callosum who have only very subtle or no obvious defects in higher cognitive functions. This probably doesn't mean that we don't need a corpus callosum, but rather that if substantial bits of our nervous system are missing, the remarkable adaptability of the brain allows it to compensate.

Other important connections of the cerebral cortex occur between cortical regions, known as corticocortical pathways. Sometimes these connections are mere millimetres in length, and in other cases they can be long axonal projections stretching centimetres. Corticocortical pathways can also be organised in an apparent hierarchy related to their levels of sophistication in information processing. In regions of cortices that correspond to processing of sensory information, for example, there may be a hierarchy in terms of the complexity of information that is processed from primary to secondary, then tertiary, and so on. Each level also includes a substantial connection back to local cortex as well as the previous level. This suggests that neurons process information in a 'feed-forward' direction, but there is also a feed-back loop connecting back to these neurons as well as to the previous step, or even to deeper previous steps. We will return to the importance of this pattern of 'recurrent' connectivity in later chapters.

Brain plasticity

While we know that there are critical developmental periods for the establishment of these sensory and motor maps, the human brain retains a capacity to shift the borders of mapped sensation and motor function to compensate for injury. For example, if I were to lose a finger in some unlikely but terrible typing accident, then my motor and somatosensory maps would reorganise to accommodate. There are limits on these shifts in maps in adult brains, which may explain why some people experience the sensation of phantom limbs following an amputation. These maps clearly have a neuroanatomic basis: they are formed during development by specific axonal pathways, but then appear to be functionally malleable.

There are also regions of cortex that appear to be responsive to multiple sensory pathways –– often referred to as the association cortex – with substantial and reciprocal connections between them and perhaps also including overlapping topographic maps. These massively interconnected regions, which bring together a range of sensory information, may be closer to the action in terms of our flowing mental awareness of the external world.

In this chapter I have provided a whistlestop tour of the cerebral cortex, but it is important to note that it does not work alone. There are substantial pathways into the cortex related to various sensory modalities, and there are pathways out for action, including the control of muscles.

There are also loops of connectivity from the cerebral cortex involving subcortical masses of neurons, arranged in 'nuclei', that are linked with a myriad of functions such as awareness, cognitive processing, sleep/wake cycles, processing of emotions, memory formation and motor control. Many of these are well preserved across species with brains.

So much for the brain's structure and connectivity, which are increasingly well understood. The next chapter will explore how these physical features might give rise to something much more mysterious: subjective experience.

Chapter 3 Some established ideas about how brains work, and a newish one

In the first chapter, I imagined how unimpressed an observer might be if they were to see the fatty pink blob that comprises 'you'. As we've seen, an extraordinary degree of organisation becomes apparent when we look more closely at the brain. But how does that intricate structure create the ability to feel, sense and move, not to mention consciousness?

There are all sorts of ideas about how brains perform their spectrum of neural functions. Broadly, there is a view that the huge number of individual brain areas and circuits work together when required as a 'coalition' of neural control systems. This view stems from three main sources: studying patients with localised brain damage, recording electrical activity directly during brain surgery, and, more recently, using brain imaging to track subtle changes in blood flow as a measure of brain activity, as well as other imaging approaches as proxies of brain activity.

From these approaches, we now know there is a brain area that seems to correspond to when your foot feels something, for example, and another when it needs to move, and that there are other areas for understanding language, or regions for producing language. We have also learned a great deal about the localisation of brain function from studies of non-human experimental models. Historically, this has involved destroying very small regions of brain and looking for functional deficits. Most recently, 'optogenetic' methods can switch specific nerve cells of the brain on or off, using light to control neurons that have been genetically induced to express light-sensitive channels in their outer membrane.

New methods for imaging have revealed the pathways of axons in human brains, helping to show how areas of the brain may be functionally interrelated for particular sensory experiences or mental intentions.

In many ways, we now understand the anatomy and physiology of neural systems for controlling relatively simple functions, for which we usually don't have any conscious experience, such as breathing, heart rate, and motor reflexes. We also know a lot about nervous systems embedded in organs, such as the enteric nervous system in the gastrointestinal system that controls the movement of food along its length. It's not that these systems are completely understood, but they give rise to scientific questions that are much simpler to study than conscious perception. Scientists still often can't agree on what conscious perception actually is, let alone how our brains create it.

Not to be put off by this trickiest of neuroscientific questions, this chapter will focus on ideas that relate to how brains may generate that pinnacle of human existence – conscious experience. There are dozens of ideas, hypotheses and theories that seek to explain the neural/philosophical basis of consciousness, and this chapter will cover some of the leading contemporary proposals, with suggestions for further reading at the end. These concepts tend not to focus on the 'why' of consciousness – a question that may point to new insights into how our brains function – so this chapter will conclude with a new idea about the evolutionary purpose of consciousness.

A common view of how brains work to produce sensory perception is that sensory information is broken down into small elements that are progressively re-assembled by hierarchical processing steps into cells and circuits that are increasingly sophisticated in their sensory complexity. This hierarchical processing is not only a feed-forward process, from lower to higher brain centres in terms of complexity, but is also influenced by very local 'recurrent' connections as well as feed-back connectivity between cortical regions at each level. This offers two layers of feed-back to a cortical region, which increases the sophistication of information processing and may contribute to the most mysterious of brain functions, conscious experience.

Association areas of cortex then combine information from different cortical processing areas, particularly between sensory modalities, and also those involved in preparing for intentional action, such as motor activity. Francis Crick and Christof Koch were among the first to champion the concept that the synchronised activity of disparate parts of the cerebral cortex may constitute a 'neural correlate of consciousness'. The assumption is that processed sensory information enters into these synchronised networks to produce a conscious percept.

Another interesting theory related to conscious perception has its origins in Bernard Baar's Global Workspace Theory. This suggests that there is a central network of cortical neurons that brings together brain functions such as attention, intention and working memory and represents our conscious mental state. This central network can then be influenced by, or can access, otherwise isolated unconscious specialised processor networks that subserve specific functions, such as long term memory. In this way, the central Global Workspace is a kind of scratchpad or blackboard where the work of creating a mental state is done, which can sample or be influenced by unconscious work being done elsewhere in the brain. The Global Workspace may cause an 'ignition' in other brain regions that bring information into the blackboard workspace. Long excitatory corticocortical connections from the Global Workspace to other parts of the brain allow this rapid engagement with processing networks to occur. In addition, inhibitory processes are also likely important to dampen the influence on non-salient processing units.

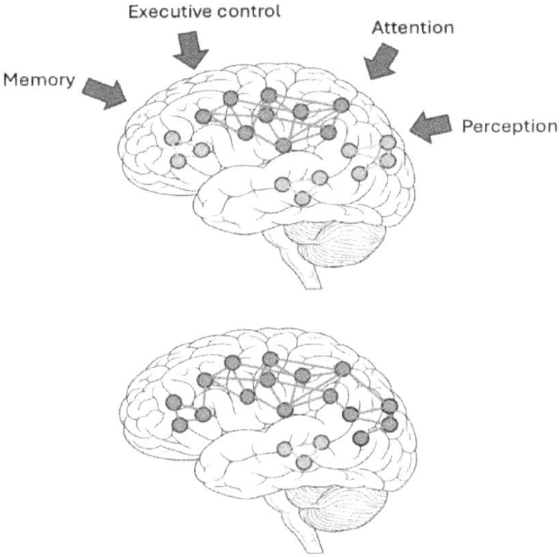

Global Workspace Theory. Information is received by the brain into a multitude of processing networks (arrows indicate examples). When they reach a threshold, they are incorporated into the Global Workspace network and a conscious event is 'ignited'.

Another compelling view of conscious mental states comes from the Theory of Neuronal Group Selection developed by Gerald Edelman and built upon through a number of collaborations with scientists in this field. This theory holds that Darwinian natural selection is a general organising principle during the development and operation of the brain. Initially, all neurons compete to form good connections during brain development. We generate more neurons and connections than we can use, so competition for connectivity produces winner neurons and loser neurons. The latter either self-destruct or their connections are 'pruned', which means they have less influence. The competition continues to a less lethal degree throughout life, with those neurons and networks that are actively stimulated having the edge in forming and maintaining synaptic connections with other neurons.

The magic sauce of this neural Darwinism is the concept of 're-entrant signalling'. This relates to the reciprocal connections that are found between areas of the brain – not only– from within the modules that comprise the cortex, such as through recurrent local connections, but also between the cortex and the thalamus. In this model, any activity and/or directionality of information flow also brings synchronous feed-back through re-entrant pathways, which may be essential for generating a stable percept or mental state. The use of these feed-back/re-entrant loops brings greater sophistication to the processing of information and may help to bind a range of internal networks to produce the final percept, such as the sight, sound, smell and touch of an object.

Overall, many of these models present a general view of the brain as a sensory processing unit that binds information together through connectivity within and between the two hemispheres. Perceptual binding and conscious awareness are poorly understood but are assumed to relate to the sophistication of circuitry, including connectivity, modular cortical processing and synchronicity of activity. As we will see in later chapters, it is important to note that a substantial amount of neuronal activity and connectivity that goes on in the cerebral cortex that not be directly related to the conscious experience.

A certain degree of passivity in brain function is assumed in many of these models – the creation of a conscious percept is something that happens in the brain rather than a property that the brain drives. Conscious perception may be closely related to the master control of attention, which may be engaged, for example, by collective binding of activity of neurons at particular frequencies of activation driven from subcortical input and/or through feed-back loops of connectivity.

The hard problem remains

There are also some significant challenges with extrapolating from the neural correlates of consciousness to consciousness itself. The models may not actually tell us how these circuits produce the subjective experience of conscious mental states. We can understand, albeit imperfectly, how nerve cells connect with each other and how they operate, but that may not provide insight into, for example, how we see a plastic duck, note that it is yellow, or hear a squeak when we squeeze it.

Neuroscientists have made significant progress with developing algorithms and computational models that may explain how nerve cells interact with each other, particularly how relatively simple nervous system networks operate. As noted earlier, substantial advances have also been made in understanding behaviour in simple animal models, and how bodily functions such as respiration, blood pressure and peristalsis are regulated.

An emerging view of how the brain may work, known as predictive coding, is that the nervous system continually makes predictions about the external world, possibly by way of feed-back pathways at different levels of processing and/or re-entrant connectivity. Further local processing at these different levels may help with finessing the final percept. This is best illustrated in a range of sensory illusions, which indicate that top-down processing can alter perception. It has been argued that this indicates that what we consider our 'conscious mental state' is a balance between sensory input and constant feed-back from other brain systems that creates an internal 'simulation' of the external world. There is some solid research to support the idea that our conscious state is a simulation, or a set of brain-produced and controlled hallucinations. But this concept still often assumes that there is a passivity of experience: that conscious awareness is a little 'you', a homunculus inside the brain who is looking at the simulation.

The philosopher Daniel Dennett has criticised this view of conscious mental states as being a 'Cartesian Theatre' (named after Descartes's philosophy of mind-matter duality), in which there is a relatively knowable set of neural systems responsible for functions such as sensation, but an unknowable neural entity looking on and calling the shots. In Dennett's view, brains work by receiving external stimuli, integrating them and then producing some form of response/behaviour, without any need for a theatre of the mind. This approach to mental states has been associated with computational views of brain function. It supports a proposition that all of these neural pathways are ultimately understandable in terms of how information is moved around the brain. In theory, this could be simulated by artificial intelligence approaches or by complex sets of transistors and connectors. That is, if we were to know all there is to know about how the human brain is organised and functions, we could then create an artificial simulation that would give rise to what we understand to be conscious mental states.

From a neuroscience point of view, this is an appealing idea in that it supports an information flow view of brain organisation. But it is somewhat unsatisfying in terms of how we feel our minds work. With the artificial intelligence parallel in mind, large language models currently exist that make it possible to meaningfully process complex information and present the appearance of human-like understanding, but many philosophers of the mind note that this is an unconvincing simulacrum in terms of the awareness of processing and intentionality of action that characterise conscious human mental states.

There is an abundance of books and papers on the philosophy of mind that attempt to keep pace with the fast-moving world of neuroscience research. The prevailing view is that neuroscience will explain conscious mental states, eventually. However, it is important to note the contrary view

– that brains are so complex that the human brain will struggle to understand how the human brain works, so it is probably better to work on other scientific questions at a lower level of explanatory ambition. Many also believe that our nervous system, even in all of its complexity and ability, may never be sufficient for explaining conscious mental states.

John Eccles, an Australian neuroscientist who won a Nobel Prize in 1963 for discovering how nerve cells communicate with each other, was in the latter camp. Eccles, often working with the philosopher Karl Popper, suggested that there are different states of existence or 'worlds', with mental awareness occupying one world and the brain existing in another, with interactions ensuring that both were functionally interlinked. This 'dualist-interactionism' was a relatively fresh take on a long tradition claiming that conscious experiences cannot be adequately explained by brain function alone. However, the consensus view among neuroscientists is materialist and holds that conscious states are rooted in neurobiological processes.

In more recent times, some leading thinkers in this area have suggested that our conscious states are fundamental and potentially irreducible, like time, space and matter, and a relatively inscrutable aspect of reality that may even occupy another dimension. This does seem to align with the Eccles-Popper proposition that consciousness resides in another dimension or 'world', sailing close to philosophical equivalence with dualism.

Sometimes it feels that it may just be easier to align with the 'mysterians' who hold that phenomena such as consciousness are beyond human understanding. There is something comforting in that approach, as it removes the need to tackle and solve these mental states as a scientific problem. Others would argue that arguments about consciousness are a waste of time and mental effort, as we are undone by the language we use to describe mental states.

Is consciousness a quantum phenomenon?

There is no getting away from the amazing nature of the brain, including its ability to undertake a multitude of tasks quietly and effectively, and to produce marvels such as consciousness. The latter joins the ranks of wonderful yet 'spooky' features of the universe including quantum physics, the nature of matter, and time and space. The scientific race has been to understand all of these marvels at their fundamental level. Despite some tremendous insights, final understandings have been elusive.

Some brain scientists have been suggesting that consciousness may be as fundamental to existence – and as inexplicable – as features such as time and space. Others consider that very strange, conceptually tricky areas such as quantum physics and consciousness may be linked. For example, Roger Penrose and Stuart Hameroff have proposed that quantum processing may occur inside neurons, in structures such as microtubules, to create synchrony across neural systems, with the synchrony equating to the conscious state. Related to this concept is the view that regular computational processes may not be adequate to explain consciousness, and– that in fact a new physics is needed to account for how neuronal activity produce this phenomenon.

Others speculate that advances in artificial intelligence will lead to a machine version of human conscious states, although it seems implausible to equate machine intelligence, no matter how powerful, with the mushy blood and guts consciousness that humans blithely enjoy.

Subjective nature of consciousness

A substantial amount of academic interest in consciousness relates to the subjective experience of the phenomenon. Classically, this is framed as the question of what it is like to see the colour red, taste an orange, feel sad, smell a rose, experience the sensation of wind on one's face, etc. While many of these internal mental states can be described with similes and metaphors, it is notoriously difficult to describe the actual feel of that brain state, as we can't actually feel the brain itself working to create that experience. We can often infer what a conscious experience is like in other people by reflecting on our own experience of that state. Assuming, of course, that everyone else is as conscious as me!

These subjective states are distinct, such that a rose does not smell 'red', the wind on your face doesn't feel 'red', when you feel sad it doesn't feel like 'red'. That doesn't mean that a particular sensation can't colour another, such as when, for example, aromas may moderate an emotional state. We can also see where brains can muddle these filters. In the condition known as synesthesia, sensory modalities become unusually linked with each other. A person with synesthesia may report that letters, numbers, sounds or other sensory sensations may be associated with particular colours. However, they are not disturbed by this experience, it's just how their mental states work.

Why a Brain?

We could try a different approach by not asking, 'how does the brain produce consciousness?' but instead asking a more basic question: 'What is the brain actually for?'

You are not aware right now of the electrical storms and neurochemicals passing between your neurons, the constant recalibration of your blood pressure, the molecular machinery releasing and recycling neurotransmitters across trillions of synapses. You are not in the least bit aware of your brain doing any of this. You are aware of the chair you are sitting in, this book, the sounds from the next room, whatever is gently circling in your thoughts. You experience the world outside your skull but not the organ inside it.

In this respect, the brain's most clever trick may not be to process the world, but to place you in it.

The brain's core evolutionary task may be to generate a continuously updated account of where you are, what is happening around you, and what you should do next. The brain does something remarkable to achieve this. Rather than passively receiving streams of sensory data and then assembling them into a picture, it works the other way around. The brain actively generates a model of you in the world and uses incoming sensory signals to refine and correct that model. It is not just a receiver of information; it is a projector of information.

What we experience as our mental state – our conscious awareness of being here, now, in this place, doing this thing – is the brain's outward projection of itself into the environment, into different kinds of spaces. The projection is not something the brain produces and then displays to an inner audience. There is no inner audience. The projection itself is the experience. It is you. This concept can be described as the Projection Space Theory.

As an example, consider what happens when you reach out and pick up a cup of coffee. Your fingers curl around ceramic that is warm and smooth. The warmth and the texture are not 'in' your fingers. They are experiences generated by your brain, which has mapped your hand into external space and is continuously refreshing that map based on what the nerve endings in your fingertips are reporting back. The projection extends to the cup: it reaches out of your skull and settles on an object in the world. When you set the cup down and reach for a pen, the projection immediately accommodates that new reality. The weight, the smoothness, the narrow cylinder between thumb and forefinger. You do not notice this happening because it is happening perfectly. It is the seamless projection of your brain colonising the world around it.

This is also why reading a novel works as well as it does. When you are deep in a story, the outside world falls away and you are transported into the world the author has constructed for you. Your brain is doing the same thing it does with the car, or the coffee cup, or the room. It is generating a continuously updated account of you in a space. The space just happens, in this case, to be imaginary.

In this framework, the projection the brain creates is not a representation of your experience. It is the experience. There is no additional step. There is no inner observer sitting somewhere behind your eyes, receiving the projection and interpreting it. This way of thinking about the brain has a natural grounding in biology. Evolution does not produce organs for the sake of it. Every structure, every circuit, every system we find in the nervous system is there because it improved survival and reproduction. The question of why the brain generates conscious experience – a question that has engaged philosophers for centuries – looks different if we approach it from this angle. The brain generates the projection because an organism that is seamlessly present in its environment – that experiences itself as inhabiting that environment rather than processing it from a distance – is better at navigating, foraging, avoiding predators, forming alliances and raising offspring than an organism that does not.

The experience of being here is not a mysterious by-product of neural computation. It is what neural computation is for.

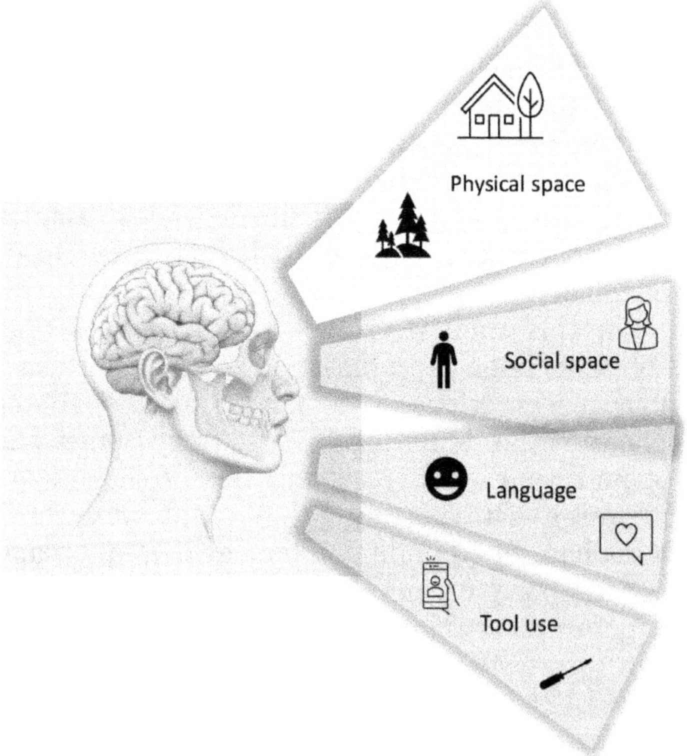

The brain is projecting itself out into the external environment. A number of different 'spaces' are colonised by your brain's activity, and the projection is the basis of your conscious experience of this engagement.

This also helps make sense of what happens when the projection fails. We are familiar with the devastating losses that accompany conditions such as Alzheimer's disease: the erosion of memory, language, and the ability to recognise familiar faces. But there is something else that is harder to articulate, which is the loss of presence. People in the later stages of dementia often seem to have difficulty being anywhere. They are not simply confused about where they are, they seem to have lost the capacity to be solidly located in a space at all. The disease has damaged the very machinery that places a person in the world, so the world itself becomes unstable, indistinct, and eventually unreachable. We will return to this in Chapter 8.

By the same logic, the distressing hallucinations of psychosis and the disturbed sense of self that accompanies severe depression can be understood as failures of the projection at different points in its machinery. The brain is still running, but the account it is generating of the organism in the world has gone wrong in particular ways.

The sensation of moving through time

One final feature of the projection deserves emphasis before we look more closely at how the brain achieves it. The projection is not a series of snapshots. It is a continuous flow, a rolling account that carries temporal continuity, so that the present moment is always experienced as emerging from an immediate past and moving toward a near future. This is why we experience ourselves as moving through time rather than simply being at a point in time.

Memory reaches back along this timeline, and imagination reaches forward. When we recall yesterday's conversation or picture tomorrow's journey, we are doing something structurally similar to what the brain does in every moment of waking life: projecting into a space. The difference is that the spaces are past or future rather than present. The same machinery is at work. This is a point we will develop further when we consider memory and imagination in Chapter 6.

For now, let us hold on to a single image. Your brain is a bodily organ sitting in a dark, enclosed skull. It has never seen the world it so vividly presents to you. It has access only to electrical signals arriving along nerve fibres, from which it constructs – continuously, effortlessly, and for the most part flawlessly – the experience of being a person in a place, moving through time, connected to other people, inhabiting a body. That construction is the projection. And the projection is what it is to be conscious.

In the next chapter, we will look at how the brain is organised to produce the projection, what we know about the circuits and systems involved, and what Projection Space Theory explains that other approaches to consciousness struggle to account for.

Bibliography

Baars BJ (1988) A Cognitive Theory of Consciousness. Cambridge University Press.

Dennett DC (1993) Consciousness Explained. Penguin.

Dennett DC (2018) Facing up to the hard question of consciousness. *Philosophical Transactions B*. 3737: 20170342.

Edelman GM, Gally JA (2013) Reentry: a key mechanism for integration of brain function. *Frontiers in Integrative Neuroscience* 7: 63.

Popper KR, Eccles JC (1977) The Self and Its Brain. Springer International.

Seth A (2021) Being You: A New Science of Consciousness. Faber & Faber Ltd.

Seth AK, Bayne T (2022) Theories of consciousness. *Nature Reviews Neuroscience* 23: 439-452.

Chapter 4 The brain projects us into the external world

Brains do not just do one thing at a time, they perform a multitude of functions simultaneously, not all of which we have constant conscious access to. Broadly, there has been a tendency to conceive of brains as comprised of a number of different co-evolved systems, many of which run in parallel with variable amounts of crosstalk and coalition, and that some systems are more closely linked to functions such as conscious perception than others.

It is important to consider the lived reality of the brain as a body organ. A brain exists in a skull – it is in a dark, very warm place that, as a consequence of evolutionary pressure and selection, has not evolved to be too overtly concerned about its very local intracranial environment or its state. In other words, there is very little mental representation of our brain within the skull. We don't feel like a blob of fat and protein, or a complex hotch-potch of neuronal connections, in an enclosed space, our skull prison. Instead, we experience the brain's inner working mainly through subjective experience, such as sensory perception, a flow of an internal narrative, and a variety of impulses and emotional states.

As introduced in the previous chapter, the main purpose of a brain, shaped by thousands of years of evolution, may be to place us into, and guide our response to, the external world. Our brains don't operate as though we live and experience mental states in a bony box; instead, the experience of this neural machine is to make us present in the world outside of the cranium. Hence, much of the workings of the brain are about placing or *projecting* us into external spaces, where it works to create a constantly refreshed account of us in the outer world so that we can meaningfully interact with our environment. The projection works as a set of overlapping maps of representation of the external world. The maps are multitudinous and can include a variety of sensory modalities, as well as our physical position and relationships and interactions with the material environment and other organisms.

The brain, locked away in the skull, does not just receive information through various sensory modalities, but actively seeks 'presence' in the external space: the machinery of the brain projects a representation of its workings into spaces beyond its physical location. To place itself into a construct of the external spaces outside the skull, the brain creates a running percept based on information it receives. The sensory systems that process this information do not lead directly to a percept, rather the brain is constantly creating a projection that efficiently draws on the hierarchy of centres underlying sensory systems, as well as some predictive computation. Hence, the brain is not just information processing, it is ***information making*** to place us in its own representation of the external world. Generating these features would then represent the conscious perception across modalities. As a product of neural construction, consciousness resembles the idea of a 'controlled hallucination' proposed by Anil Seth and others.

Projection Space Theory proposes that the brain continually builds models of how things may be working in the world external to the organ's confines. In this way, substantial regions of the brain are not involved in passively processing sensory information for yet higher levels of integration, but rather are engaged in creating the perception based on what sensory inputs it can sample compared with a set of predictions. The brain is running a mapped representation of the external world so that it is projected into, and feels as though it exists in, these external environments, which could be physical, social, or even simulated, including past and future space states.

If I hear a voice in a particular direction, for example, that experience is enabled by my ears and auditory system but this input fits into a running projection of my physical position as well as the external auditory space. An itch at the back of my neck does not exist other than as a projection of the brain that is stimulated by the sensory input mapped to a spot on my skin which also assists a scratch. A lover's familiar touch on the hand as an isolated event is neither actually perceived or felt by the hand, but rather as a projection of the hand by the brain, overlaid with filters of meaning. If another strokes our hands, the filters may tap into memory and imbue the projection with warm tenderness if it's a loved one, or some degree of apprehension if the person touching is a stranger. Hence, the neural filters provide an important value-add to the process or projecting us into various paces.

The projection is 'you'

This differs somewhat from the idea that brains create simulations or 'controlled hallucinations', as this would still necessitate some kind of homunculus within the brain viewing a simulation and guiding physical action. Where Projection Space Theory differs is that the brain essentially creates a projection of you into the external world, and the projection is the conscious mental state – it is in fact 'you' – which negates the need for a further entity or neural construct observing a simulation.

These projection maps between the organism and the external world, including how we interact with technology, are also flexible and extendable. We can return to the earlier example of driving of a car. While driving, substantial sensory information is coming in via the visual system, but we also have a sense of the position of our body and the car in space, and the projection includes planning for the direction of driving to arrive at a required destination. The projection also includes the feel of the car on the road – it's as if the map of our body in the external world also extends to the vehicle – we can 'feel' the connection between the tyres and the road, and we include in this projection how the engine is running, feed-back through the steering wheel, positioning in a lane, etc. The brain then works not just for the body, but for a body ***plus*** a car, adding in sensory inputs that are provided by the technology working in a space. While you drive, the brain extends into the technology and projects into the outside world as the car. The brain does this effortlessly as it is core brain business, and extension of the projection enables the successful use of this technology towards an intended outcome.

Conscious perception is also likely linked to this projection of the brain to place us outside of the skull. Perception then isn't a result of some central mysterious homunculus looking at various bits of sensory data as they come in, rather the brain creates a flowing account of us in the external environment, influenced by available sensory data as well as internal predictions of current and future states. The magic in this is that the experience seems seamless – we don't need to make much overt mental effort in the formal flow of the projection. What we refer to as conscious experience of our mental state may be the same as the running projection of the brain. The flowing manufacture of a projection into external spaces also creates the conscious perception of the flow of time. The projection is constantly created, with the most recent projections lingering to provide a sense of movement through time, as well as anchoring to most recent past actions. The brain projection into external spaces drives a binding of available information, and such spaces can be real, vicarious, emotional and social. For example, reading a novel or watching a movie we project the fictional characters, their actions and relationships as if we were actually there.

Projection into social space

Projection space mapping also provides a mechanism for the full range of interactions we have with the external environment, including acting as the prism for inter-individual differences. In other words, our personality is part of the 'how' in which the brain projects ourselves into the space around us. The spaces we project into can also overlap with other brains' spaces, and this is where language and other forms of communication provide a way for brains to interact with each other without any need for direct exposure to the goriness of the organs themselves. Our brains seek to interact with each other in this virtual external space and we interpret the sounds created by another human organism as speech, overlaid with a variety of filters related to memory, emotion and context.

Attention is a way that the neural machinery of projection focuses to sample a subset of the available information, processing it and creating the projection. Wandering thoughts and dreams are constructions of the projection system when it is not actively engaged. Left to its own devices, somewhat detached from focused sensory sampling, neural projection machinery creates coherent and semi-coherent flows of mental activity. A particular set of neural connections within the cerebral cortex is associated with this constant flow of the brain 'at rest', termed the default mode network, that conversely appears to be deactivated when the brain is focused on a specific mental task.

Hallucinations are often a feature of both neurodegenerative disease and psychiatric conditions, and these likely represent the engagement of projection or constructivist parts of the brain. For example, in Lewy body dementia, detailed and well-formed visual hallucinations of people or animals can occur, but the person with this condition can often find these untroubling while at the same time being aware that they are not real. Conversely, auditory hallucinations in schizophrenia may take the form of internal non-self monologues with compelling messages, where there may be little insight that these are not 'real'. While these conditions have substantially different pathophysiological underpinnings, they do show that the brain is capable of producing a conscious experience irrespective of actual sensory input. Likewise, in dreams we also have limited control of a rolling experience that has features of waking consciousness. Dreams are a remarkable example of brains constructing a vivid story likely out of range of disparate neural traces, mimicking the process of projection when awake.

How does a brain achieve this projection mapping? We already know that the exterior world and even our own bodies are mapped to particular areas of the brain. These maps likely originate in developmental pathways and are guided in their formation by the activity of particular genes, but they also appear to develop in response to the competition of sensory pathways into the cerebral cortex. We can reconceive many of these hierarchical sensory processing pathways as not just being receptive to 'incoming' information, but also as the generators of the brain projection into these spaces. Both the feed-back pathways at different levels of these hierarchies, as well as lateral pathways and corticocortical connections, likely help to coordinate a seamless projection, potentially with some level of top-down manipulation linked to attention. Further sets of corticocortical connections help to maintain a synthesis between sensory modalities as well as with other cortical circuits that add a filter or context to the running projection, such as memory or emotion. There may also be a role for a master conductor: a set of neural pathways that helps to 'ignite' and bind together the activity of cortical circuits, such as has been proposed by the Global Workspace Theory. Furthermore, there may also be other circuits responsible for 'keeping time', ensuring that the activities of different areas are synchronised, potentially by subcortical inputs such as from the thalamus, or by the action of recurrent loops of connectivity and activity within the cortex itself.

We've seen how Projection Space Theory helps to avoid the philosophical conundrum of the 'Cartesian Theatre' in which a homunculus or 'little you' inside the brain observes a screen on which conscious percepts are displayed. The following chapter considers the cellular architecture of the cerebral cortex with a view to how this may support the brain's projection into the exterior world.

Chapter 5 The neural architecture of conscious experience

The cerebral cortex appears as a pinkish bloody mass with grooves and bumps, and is really a sheet of cells that is scrunched up to fit inside a skull. For those of us in research who have had the honour of retrieving brain donations *post mortem*, the organ is not great to look at 'in the flesh' for all of its gory features, but it is also always an awe-inspiring experience to realise how a person and all of their remarkable attributes and functions are bound up in its organisation.

How does this unpromising mass of tissue give rise to the wonder of human consciousness?

A very tricky problem for reconciling any theory of consciousness with what we know about the cellular composition of the brain is how various elements interact with each other. Broadly, there are two types of cells that are present in brains:

- Nerve cells which process information and communicate with other cells, mostly other nerve cells, via synaptic transmission.
- Various types of 'glial' cells that have a myriad of functions which include supporting nerve cells metabolically and in synaptic transmission, sculpting or pruning synaptic connections, and protecting the brain from infection and damage.

It has been estimated that there are about 80 billion nerve cells in the brain, and it would be a proportion of these that are either directly or less directly involved in generating a conscious experience. The two hemispheres of the cerebral cortex are generally thought to have the most important role in generating a conscious experience, but that these are substantially interconnected with other brain regions, particularly in the receipt of sensory information and the generation of motor activity and interactions with other bodily systems.

Postmortem studies of people who have had disastrous things happen to their brain support the concept that conscious experience relies heavily on the top-most parts or mantle of our central nervous system, the cerebral cortices. When we look closely at this sheet of cells that corresponds to the 'grey matter' of the cerebral cortex in cross-section, we can see that, at the level of organisation of nerve cells, there is a highly modular arrangement of these neurons that extends across cortical regions, whether are largely responsible for motor or sensory functions, or the more mysterious regions that may be involved in functions such as the association between sensory regions, laying down memories, or planning. There are neurons with tear drop or pyramidal shaped cell bodies, often referred to as pyramidal cells, that have projections that extend out of these cortical modules but may also have side connections back to local microcircuitry. Then there is a large group of nerve cells with differing shapes, which we can refer to a non-pyramidal cells, which tend to have more local connections within modules, but also lateral projections between modules.

Pyramidal cells are the most numerous nerve cells in the cerebral cortex. They have a long main 'apical' dendrite that extends perpendicularly from the apex of the conical cell body through the sheet of cells of the grey matter, and a complex, tree-like set of branched dendrites through which they receive input from thousands to tens of thousands of other nerve cells. The balance of excitation and inhibition of the pyramidal cells is then summed at any given moment to trigger or inhibit an electrical signal, which travels down a specialised process of the neuron called the axon that usually emerges from around the base of the cell body. The axon may project to other brain regions, including nearby modules as well as within a hemisphere or across to the other side of the cerebral cortex or to brain regions below the hemisphere including the midbrain, brain stem or spinal cord. Between cortical regions involved in processing sensory information, many of these pyramidal cells may have receptive fields and activity that are particularly sensitive to features of sensory data. In visual regions, for example, this may be to the edges or orientation of structures. Where such information is processed across brain regions in a hierarchical fashion – with nerve cells becoming more responsive to more complex information – there are connections that underpin this pathway, known as feed-forward connections, and recursive connections back to prior levels, known as feed-back connections.

Pyramidal cells appear to be the main 'effectors' of cortical processing, along with multipolar cells located in the deeper cortical grey matter, as their axons typically end in synaptic terminals that will release excitatory neurotransmitters. Their job seems to be to switch on nerve cells that they contact, or, at least, increase the chances of the next nerve cells firing off an action potential down their axon. These pyramidal cell axons also have a curious feature that may play a vitally important role. They have a side branch that emerges not far along their length and bends back to connect with targets within or close to the module in which they emerge – another form of recurrent connectivity. This is an interesting dimension of cortical organisation as it presumably provides a form of recurrent feed-back to the local module on the activity it impacts at more distant sites.

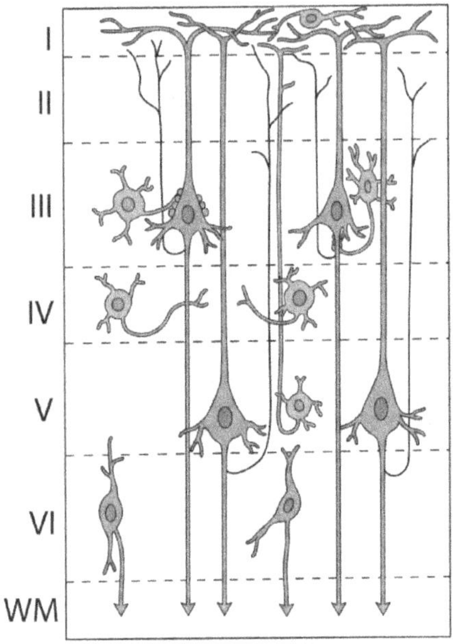

Diagram of the 6 layers of the neocortex and underlying white matter (WM), including excitatory neurons (blue) and locally projecting 'interneurons', most of which are inhibitory. The diagram includes examples of axonal projections and connections. The pyramidal and multipolar cells project away from the region, including across the corpus callosum between the two hemispheres, to other neocortical areas and subcortically. Pyramidal cells will also include a branch of their axon that forms a local projection – to form 'recurrent 'connections.

The axons of the non-pyramidal cells in layers II to V tend to be more localised to individual modules, or provide lateral connections between modules. This includes axons that give rise to both inhibitory or excitatory transmission – encouraging or discouraging other nerve cells to trigger an action potential. The role of these non-pyramidal cells has mostly been considered to be 'modulatory' to the effector pyramidal cells, but they also likely have important roles in coordinating nerve cell activity between modules and to more distant sites through lateral connections.

Overall, the cerebral cortex, like other parts of the brain, likely acts as an ongoing symphony of many cellular instruments without a central conductor but acting in coordinated fashion based on their connectivity and particular cellular properties. Many of these pathways are developed and set through development, where neurons behave in a Darwinian fashion to extend their processes and seek useful connections with other nerve cells. Those neurons that make a synapse with other neurons are 'successful' and survive, whereas developing neurons that miss out on functional connectivity may be reduced in their connection or even commit cellular suicide through a process known as apoptosis. Non-neuronal cells such as microglia may assist in the sculpting process of the branching of neuronal projections as well as in the formation or elimination of synapses.

There are relatively intense periods of brain development when nerve cells are born, potentially migrate to their positions in the brain, and send out processes looking for the right nerve cells to connect to. In humans, this developmental period extends up to and through adolescence into early adulthood. Brain imaging studies have shown that early adulthood is associated with atrophy, or the loss of brain substance, which likely corresponds to the sculpting and fine-tuning of connections towards the 'adult' form of the brain.

This radical adaptability of neurons and their related structures may not completely stop in adulthood. Studies across a range of experimental models have shown that synapses continue to be formed and eliminated in 'mature' brains, and that this modification in the functional relationship between neurons relates strongly to either the stimulation or removal of activity within pathways of connectivity. In this regard, there may be particular synaptic connections that do not change, or remain 'stable', over time, whereas adaptability of the nervous system may be linked to those synapses that turn over, or are created, modified or eliminated in response to patterns of stimulation. This remarkable aspect of brain plasticity may also be important for our capacity to learn and form memories, and may be harnessed through forms of cognitive stimulation to stave off the impacts of brain degeneration, as we will see in later chapters.

A further aspect of plasticity in the brain is the existence of reservoirs of adult stem cells within or close to the brain, with these cells able to become a range of both non-neuronal and neuronal cell types. We don't have a complete understanding of why these pockets of stem cells are maintained throughout life, but they may relate to areas of the brain where cellular turnover is functionally important. As an example, neural stem cells are found in a particular region of the hippocampus, the dentate gyrus, where they possibly have a role in the process of encoding new memories.

Refreshing neurons in this region may ensure that this neural processing step is maintained throughout life.

Effector neurons as projectors

Given these cellular features of the brain, as well as how it is generally organised (as explored in Chapter 2), how would this setup serve to project us into the external environment, and what may this say about conscious experience? The nerve cell type that is likely to be important in this regard is the 'effector neurons', the pyramidal cells that connect between different cortical regions either in feed-forward, feed-back or associative patterns of connectivity. As noted earlier, many of these pyramidal cells also have a curious side branch from their main axon that provides a projection back to surrounding local circuitry – a form of recurrent local connectivity.

If we consider the canonical model of visual information processing, feed-forward pathways represent a building up of complex visual information that comes in a broken-down form from the retina and is relayed through thalamic regions such as the lateral geniculate nucleus. But a fundamental unsolved question is where does the information buck stop so that we have a visual percept? It may be that such information needs to be transferred into a set of pathways that correspond to the Global Workspace, as introduced in Chapter 3. Alternatively, it may arise from the dense associative corticocortical pathways that interlink increasingly more sophisticated areas of visual processing. Given that recalling visual information involves similar brain regions that were initially involved in processing that visual information, it would seem likely that the visual cortex is intimately involved in producing the percept as well as processing information. As we will see later, there is an ongoing argument as to whether all of the neuronal machinery in the visual processing areas at the posterior (back) of the brain are sufficient to generate visual perception, or whether this may also require sophisticated interconnections with more frontal regions of the brain.

It may be that, while hierarchies of feed-forward information processing may occur through cortical pathways that break down and reassemble complex information, either the feed-back pathways between cortical modules, and/or the local recurrent connections from the same neurons providing the feed-forward and feed-back pathways, may underlie neuronal activity that relates to ongoing conscious experience of perception. It is most likely that the perceptual element that arises from brain visual processing involves the near simultaneous activity of feed-forward, feed-back and recurrent collaterals, on the background of a range of modulatory influences, lateral connections between modules and some influence from frontal cortex.

The potentially vital role of recurrent connections is particularly attractive as it would suggest that the same regions involved in information processing may also underlie the construction of the percept itself, the projection of the brain into the external world upon receipt of incoming sensory stimulation. In addition, the local connectivity would be widely distributed throughout individual modules of the cerebral cortex, and transmission would be relatively rapid given the shorter connections, which could be important for the speedy and constantly refreshing constructions of multiple conscious perceptions over time. Such pathways would help explain the general 'flow' of consciousness, though further input from other higher cortical regions may be required for us to report on a conscious percept.

Predictions and filters

Such recurrent connectivity may help the brain anticipate or 'predict' visual information for the final perception, and, computationally, this may be how past experience influences a current subjective conscious state through information processing by nerve cells. This would indicate that the projective capacity of the brain is not passive but involves a range of predictions and filters that govern the flow of visual consciousness.

Recurrent collaterals between neurons in processing modules may also amplify or quieten visual information and could underlie synchronicity in neuronal responses to particular stimuli. At lower levels of visual information processing in the brain, this may help to bind together neuronal responses to particular features, but at higher levels of processing it may drive synchronised grouping of information that is more closely linked to perceptual objects. This emphasises the concept that conscious perception is manufactured or actively generated by neural pathways, and is not just a passive consequence of how feed-forward processing pathways are organised.

Beyond recurrent collaterals, neurons are coordinated or synchronised within a given level of the visual processing hierarchy. This may be achieved by linkages between recurrent networks, and/or through the lateral projections of interneurons in each cortical module that regulate the activation of pyramidal cells. Longer cortical connections, sometimes referred to as corticocortical connections, would bring associations of these sensory modalities together, and provide pathways and inputs for other cortical regions involved in different functions, such as attention, planning motor activity, judgement and decision making. It is interesting that detailed connectivity studies indicate that the evolution of primate brains, including human ones, is associated with an expansion of interneurons connecting to other interneurons. This suggests that the evolution of primate brains has involved an enhanced capacity for fine-tuning the activation of projection cells like pyramidal neurons.

Recurrent local collateral connections are not just a feature of sensory cortices but are a motif that occurs widely throughout the cerebral cortex. The widespread presence of recurrent local connections could potentially be the most important underpinning of the conscious perception associated with many cortical processes, or at least how information is processed locally to contribute towards the subjective experience. Higher order influences, for example from the prefrontal cortex, may also be required to 'ignite' the required level of activity in these circuits for it to become conscious, and the subsequent promulgation of sensory information into 'Global Workspace' networks would correspond to the phenomenal features of consciousness.

The illusion of the present

An intriguing feature of the processing of sensory information is that it does take time to form a conscious percept, which has important ramifications for the neural basis of consciousness. It has been estimated that it takes up to half a second, or 100 to 500 milliseconds, from a stimulus onset until a conscious perception occurs.

This raises a number of perplexing issues. Perhaps the conscious perception occurs after non-conscious neural processing, relegating the subjective experience to an epiphenomenon, or incidental side effect, without an evolutionary purpose Conversely, brain activity over around 300-400 milliseconds may be more closely related to higher cognitive functions, such as memory or attention, and complex behaviours? It has been suggested that perhaps our conscious experience is then a form of rolling memory of events that occurred less than a second previously. The brain then provides the illusion of seamless conscious experience by stitching together these memories in perceived 'real time'. This would suggest that most of the brain's work, for example in the planning of movement, may be happening in nonconscious neural pathways, and the conscious bit is just a recent record of neural activity and decisions that have already been made. We are then left wondering what the purpose of consciousness may be, as we could just imagine the brain as an unconscious information-processing automaton – a biological machine or computer – where conscious experience is not necessary for the organism to function effectively.

There are also many examples of the brain imparting a behavioural response independent of engagement with conscious perception. One such phenomenon is 'blindsight', in which substantial damage to the cerebral cortex – particularly the primary visual cortex, or V1 – leads subjects to report that they cannot 'see'. Despite this, they may still be able to react appropriately to their external environment, as evidenced by their ability to navigate around objects or even report on aspects of their surroundings, such as the emotional states of others. This has been attributed to the preservation of pathways from the retina through to other subcortical brain centres, enabling a behavioural response, and sometimes a 'feeling', independent of conscious visual perception.

There are also phenomena such as the 'attentional blink' where the subject may fail to detect a target if it is presented too quickly after previous stimulus. The concept here is that the brain may be too busy assembling a conscious percept of the first stimulus to grapple with the second stimulus. Other tricky conscious perceptions include how visual stimuli such as disks on a screen, when presented quickly as static images near each other, may appear to be moving from one position to another rather than as separate images. If you then present two separated different coloured disks in rapid succession, the disks may appear to transform in colour between the two positions. All of these are examples where the processing time needed to provide a conscious experience can lead to visual illusions. A similar phenomenon occurs when a series of static images are presented sufficiently fast that we then perceive smooth movement, such as in the case of cartoon animations, computer games and movies.

Putting the pieces together for consciousness

A neural network that involves not only feature analysis building in complexity, but also massive feed-back and cross-referencing on this process, provides one plausible explanation of the sophisticated way in which brains stitch together and navigate the external world. However, it is much less satisfying as an explanation for the phenomenon of a conscious percept. Furthermore, similar kinds of neural connections and pathways underlie the organisation of different parts of the brain, such as for auditory, somatosensory, visual and olfactory function. But there is no clear picture of how these otherwise similar bits of the brain are responsible for very different sensations. It feels like a bit of a cop-out to suggest that conscious experience may be an emergent property from such neural complexity, and that it is not reducible to an understanding of its components.

There have been a couple of prominent attempts to explain what the philosopher David Chalmers calls the 'hard problem' of consciousness. We have already encountered Global Workspace Theory (see Chapter 3). The remainder of this chapter is devoted to an exploration of a more recent theory, which has drawn considerable interest and support.

Integrated Information Theory

Integrated Information Theory (IIT), developed by Giulio Tononi and his colleagues, begins with a number of axioms or principles about the phenomenological properties of conscious experience and builds to a mathematical model that relates to the integration of information. The axioms lead to a range of postulates or rules that determine how a physical structure such as a neural system may give rise to conscious experience. In this regard, the neural basis of consciousness may rely on a number of subsystems that have defined causal effects internally within a neural circuit but are also highly integrated, with the conscious state equivalent to the 'maxima of integrated information' relating to the synchronous activity of these subsystems. A possible mechanism underlying the integrated nature of these systems and their relative ability to have a causal effect on their own function is the re-entrant features of neural pathways. In contrast, neural systems with relatively fewer complex systems may be underpowered to contribute to the maximally irreducible conceptual state that corresponds to consciousness. Hence, a neural system comprised of just feed-forward connectivity with little or no re-entrant feed-back is unlikely to contribute to a conscious state.

The mathematical model of IIT calculates the maximal amount of integrated information, referred to as phi, which can be assessed in any given system. In principle, this may not just apply to biological systems such as a brain, but also to other complex physical systems that have high levels of phi. Integrated Information Theory does have its detractors, who criticise the way it works back from phenomenology to mechanism, and the idea that other physical structures could generate some degree of phi and may therefore have a low, but not zero, amount of consciousness.

While IIT provides insight into the sufficient conditions to produce consciousness, it does not explain how the relevant brain elements produce consciousness, which is the nub of the 'hard problem'. Nevertheless, it does highlight the importance of re-entrant connectivity.

Clearly some academics in areas related to the study of consciousness are not happy with IIT. In an open letter published in 2023 with more than a hundred authors, Integrated Information Theory was labelled as "pseudoscience" with potential harmful implications for other fields, including the ethics of clinical practice with minimally conscious people through to Artificial Intelligence. They made some valid points, but the authors also claimed IIT was not a leading theory of consciousness, even though a formal or 'recognised' list of top theories does not exist.

Theories of consciousness at twenty paces

2023 was a hot year for the profile of theories of consciousness, as duelling ideas about this field received public airings. First, there was resolution of a 25-year bet between neuroscientist Christof Koch and philosopher David Chalmers. In 1998, Koch had wagered that we would know how the brain produces conscious experience by 2023, whereas Chalmers held the opposite view. 25 years later, these colleagues agreed that Chalmers had won this particular bet, and a case of wine, and that the underlying quest to solve the mystery goes on.

Additionally, 2023 saw the outcome of a possible test of the Integrated Information Theory against the Global Workspace Theory in a study funded by the Templeton Foundation. This was billed as an adversarial study involving functional brain studies that sought to localise the areas of the brain critically involved in conscious perception. The study involved several studies that used complementary methods to test specific predictions of the two theories. Proponents of the Global Workspace Theory argued that activation of the prefrontal cortex would be critical at the beginning and end of a conscious experience, while Integrated Information Theory posited that the posterior cortex was the most important brain region for the subsystems that give rise to the synchronised maximally irreducible conceptual state that underly consciousness.

Overall, the results of the study were 'mixed' and not overwhelmingly positive for either theory. On balance, there was stronger evidence for the posterior cortex having a significant role, although synchronised activity at the level predicted was not evident. In contrast, limited aspects of consciousness correlated with prefrontal cortex activity. This indicates that either the theories require more work, or that the functional read-outs of the study may have been inadequate to capture crucial elements of the neural correlate of consciousness.

Bibliography

Albantakis L, Barbosa L, Findlay G, Grasso M, Haun AM, Marshall W, Mayner WGP, Zaeemzadeh A, Boly M, Juel BE, Sasai S, Fujii K, David I, Hendren J, Lang JP, Tononi G (2023) Integrated information theory (IIT) 4.0: Formulating the properties of phenomenal existence in physical terms. *PLoS Computational Biology* 19: e1011465

Budson AE, Richman KA, Kensinger EA (2022) Consciousness as a memory system. *Cognitive and Behavioral Neurology* 35: 263-297.

Cogitate Consortium (2025) Adversarial testing of global neuronal workspace and integrated information theories of consciousness. *Nature* 62: 133-142.

Kuhn RL (2024) A landscape of consciousness: Toward a taxonomy of explanations and implications. *Progress in Biophysics and Molecular Biology* 190:28-169.

Nerzog MH, Kammer T, Scharnowski F 2016) Tine slices: what is the duration of a percept? *PLoS Biology* 14: e1002433.

Melloni L, Mudrik L, Pitts M, Bendtz K, Ferrante O, Gorska U, Hirschhorn R, Khalaf A, Kozma C, Lepauvre A, Liu L, Mazumder D, Richter D, Zhou H, Blumenfeld H, Boly M, Chalmers DJ, Devore S, Fallon F, de Lange FP, Jensen O, Kreiman G, Luo H, Panagiotaropoulos TI, Dehaene S, Koch C, Tononi G (2023) An adversarial collaboration protocol for testing contrasting predictions of global neuronal workspace and integrated information theory. *PLoS One* 18: e0268577.

Singer W (2021) Recurrent dynamics in the cerebral cortex: Integration of sensory evidence with stored knowledge. *Proceedings of the National Academy of Science* 118: e2101043118.

Tononi G, Boly M, Massimini M, Koch C (2016) Integrated information theory: from consciousness to its physical substrate. *Nature Reviews Neuroscience* 17: 450-461.

Chapter 6 The projecting brain: memory, imagination and time.

In the previous chapter, we saw how the brain may conjure consciousness from its rolling recollection of events that already occurred just a fraction of a second earlier. But how exactly does the brain encode and recall memories into our conscious state?

We know there are many types of memory, which likely have differing (or perhaps in some cases, overlapping) neural bases. The most short-lived kind of memory, which lasts a second or so and decays rapidly, is that directly associated with conscious perception. This is sometimes referred to as the 'remembered present', likely relies on recursive neuronal pathways, and may also correspond to what we experience as a conscious state. As noted in the previous chapter, it has also been postulated that conscious perception itself, by nature of its reliance on processing that takes some hundreds of milliseconds, may represent a form of immediate memory reconstituted in a seemingly continuous form so that we don't perceive the timing gap.

Then there is 'short term memory' which refers to being able to recall events or items over the last minute or so. We can hold up to 4-5 'chunks' of information in this form of memory storage. Neurobiologically, this seems to involve information bouncing around cortical circuits, particularly in the frontal and parietal lobes, but with decay in content over minutes.

Somewhat related to short term memory is 'working memory' which is an active form of memory as it typically takes some effort to keep information from decaying. This might involve, for example, repeating information over and over, in a phonological loop. It has also been proposed that there are neural systems that can maintain a 'sketchpad' of visual and spatial information. There is evidence of an 'episodic buffer' where information involving different sensory modalities as well as their chronology may be held, for ready access.

Retaining a memory trace of sensory experience over longer time intervals is referred to as long-term memory and comes in a variety of types, such as factual memory (semantic memory) and autobiographical memory (episodic memory). Long term memory likely involves changes in the nature of connections between nerve cells, and is an example of neural plasticity. There are several phases to establishing long term memory, and this includes encoding and consolidation.

Research has shown that there are particular areas of the brain that are important for generating long term memories, but these are not likely to be where the neural trace, or engram, of the memory resides. In this regard, a range of brain structures in the most medial part of the temporal lobe (the medial temporal lobe) appear to be important for the encoding of memory. Damage to these areas, either by trauma, hypoxia (oxygen deprivation) or via neurodegenerative diseases such as Alzheimer's disease, leads to significant deficits in the ability to form new declarative memories, whereas memory for events that occurred before the damage may be relatively spared.

Important medial temporal lobe areas for memory formation include the parahippocampal region as well as the hippocampus. The parahippocampal region is extensively interconnected with the sensory cortices and acts as a funnel for connections to local circuits within the hippocampus. The output of the hippocampus then returns to the parahippocampal regions, where it is subsequently propagated back into the cerebral cortex. The final memory 'trace' in the neocortex is likely represented by changes in synaptic connectivity which may have a range of potential biological bases, including alterations in the morphology of synapses or 'epigenetic' changes, such as the methylation of genes. It does seem to involve the production of proteins, which indicates that structural changes in neurons are part of this 'trace'.

There is evidence that medial temporal lobe regions may be important for the retrieval of particular kinds of memories. Furthermore, sensory areas that were initially involved in encoding may also be activated when memories are retrieved. Medial temporal lobe regions such as the hippocampus, therefore, establish an 'index' to the memory trace that help to activate its sensory aspects. Recapitulation of these neural pathways could be important for strengthening the key relationships between neurons, contributing to the consolidation of memory. Over time, it may be that neocortical regions, such as in the prefrontal cortex strengthen their role in indexing the engram for retrieval, such that retrieval of a memory trace is less dependent on medial temporal regions.

The internal replay of events can feel as though it involves many of the original sensations, but it is usually a hotchpotch of elements that we shuffle and rearrange to align with what we were seeking in the memory. Specific details may be missing or 'filled-in' and some qualia, such as taste or smell, cannot be recreated completely. However, we may recall how they made us feel at the time, which indicates some binding with other information. Interestingly, while we are not very good at internally replicating taste or smell, our memory systems can help us recognise these forms of sensory stimulation when presented with them, as well as prompting retrieval of information for other sensory modalities where these were associated with past events. In this way, a smell or taste may 'unlock' particular autobiographical memories.

The 'home' of memory

While we know that particular parts of the brain, such as the hippocampus and medial temporal cortex, may be vital for processing information and setting down memory, the 'home' of a memory trace in the brain remains a mystery. It is generally assumed that it lies in the formation or revision of synaptic contacts between nerve cells, principally in the cerebral cortex. As noted in the previous chapter, particular elements of cortical connectivity, such as recurrent collaterals, may be one way that brain function can be sculpted based on past experience. Essentially, recalling a memory involves many aspects of the integration of information that is implicated in conscious experience. Clearly, memory involves the reconstruction of a percept out of elements that comprise the engram: we are imagining the past. This view is consistent with Projection Space Theory as our neural machinery seeks to replicate a projection of events that occurred in the past, an experience of our brain state at a moment in time. This projection assembles a range of previous sensory experiences, and likely smooths out missing details to produce a flowing account. As noted earlier, this appears to involve a recapitulation of brain areas likely responsible for generating the sensory percept in the first place.

However, this projection may not be entirely reliable, as it likely involves a pastiche of previous sensory experiences that are linked together, potentially in spurious combinations and with embellishment or missing elements. Furthermore, the process of consolidation, where an engram is made more robust, also offers opportunity for errors to be made or confabulation to occur. In this regard, we should consider memory to be a version of past history that is actively and contemporaneously manufactured by neural pathways responsible for the initial projection. Memory then is clearly a way to reignite perception in a virtual sense, perhaps with the projection capacity of the brain assembling a recognisable flow of events that share features of the initial stimuli.

That is the past but how do we think about the future? Humans are also fortunate to have the capacity to develop a multisensory internal experience of the near to far future. Does this similarly rely on neural mechanisms related to the brain to create a projection? Are we projecting into the future? This appears to make sense in the context of the well-recognised predictive mechanisms that may allow the filtering of sensory experiences in the here and now.

A neuroscience of creativity

In 2020, and shortly before his death, I had the fortune to meet the design entrepreneur Gary Cleveland at his house in Launceston, Tasmania's second largest city. Gary was born in the United States and came to Launceston to manage a large textile mill which ultimately was doomed to closure following changes to trade tariffs applied to imported textiles. Gary's keen interests in design and wish to promote Tasmanian designers then led him to co-found Design Tasmania to highlight the work of local artisans. His view was that Tasmania's ancient history and natural and social environment 'slows' people and gives the space for creative thinking and, subsequently, innovative and quality design. Gary arranged for us to meet as he became interested in work we were doing on reducing the risk of dementia, as well as what research may be saying about how the brain is involved in creative thinking and especially imagination, and whether any of this may be harnessed to stave off or treat dementia.

Gary lived in a distinctive timber house that he had designed himself, on a hill overlooking Launceston's major river. The house demonstrated his dedication to design, but also showed signs of evolving features that represented his ongoing engagement in developing its form and function. Gary had experienced several health setbacks, but his engagement in topics that interested him was substantial and he had many more questions on the neuroscience of creativity and imagination than I could answer. Unlike the massive literature on the cognitive features and neurobiology of memory, relatively little research has been done on imagination. It would seem that imagination does share some common features of memory – including the ability to conjure imagery, construct scenes and populate a storied sequence, particularly with audiological and visual features, that play out over a time sequence. Imagination is clearly involved in the conceptualisation and planning of future events. It taps into our ability to assemble possibilities of a future state, likely making use of both past experience and areas of personal skill and know-how.

Gary was also interested in whether being stimulated to imagine may be a useful exercise for fortifying the brain against ageing-related neurodegenerative conditions such as Alzheimer's disease. In our conversation, I thought that this must be possible as the theory was that, particularly multimodal, cognitive stimulation may help to build resilience to dementia, not by reducing pathological damage but by providing surviving neuronal pathways with a greater number of potentially redundant connections.

Gary's interest in design and imagination taps into important brain capabilities on how we project forward in time. How the brain may generate a picture of future states has been examined in studies of spatial planning in experimental animals, which have implicated the hippocampus as having an important generative role. This has included the role of 'place cells', neurons whose activation is attuned to particular spatial fields, and may correspond to a neural basis for an internal map of the environment. In this way, brains can build a 'cognitive map' of their spatial environment, which can be tapped into to guide future behaviour such as solving navigational tasks. The discovery of place cells in the hippocampus, as well as spatially tuned 'grid cells' in the adjoining entorhinal cortex, led to a Nobel Prize for key scientists in 2014. In support of a link between memory and imagination, place cells may be important for the neural record of episodic events in terms of spatial content, but may also be involved in generating a spatial map for planning future behaviour. Recent research in experimental animals indicates that there are assemblies of these spatially aligned cells that sweep the space from just behind the animal to the space in front of it, within theta cycles of activity (6-10 Hz). These sweeps can occur to the left and right of the animal, possibly making scans and predictions for future movement through the space, and also in a goal-directed manner, for example towards food. These 'theta sweeps' help the brain form a model in space for future action. They also occur during sleep and have been hypothesised to help replay spatial events and support the formation of long-term memory. The discovery of theta sweep phenomena has been a very exciting development as it potentially provides a neurobiological model for how the brain may be making predictions about future states.

While the focus on these circuits has been on encoding spatial information, similar pathways may be involved in processing information of a 'continuous' nature that helps guide behaviour, such as in relation to odours and sound. Integration across these modalities may be particularly tuned towards goal-related behaviours – in animal models this may be about finding a path, seeking food or exploring a novel object in the environment. This would also help the brain compute a set of behaviours towards goals that it can't directly see, by referring external cues to an internal cognitive map. Likewise, this neurobiological process may help explain more generally how the brain integrates input from a variety of sources to form a neural account, which could include recall and memory. Imagination may make use of such cognitive maps to bring together and integrate information, help predict future states and orient us towards particular goals.

Binding everything together

Other research indicates that the coordination of firing patterns between groups of neurons may also be important for both conscious perception and for developing a brain corollary of future states. Electrical fields related to brain activity can have particular patterns of oscillation. As noted above, coordination of neuronal activity in the theta range may be important for neuron assemblies to make predictions about future states. There has also been substantial interest in the coordinated rhythmic activation of nerve cells in the gamma range (25-100 Hz) which may be particularly important for binding together neural processing across a range of brain areas, with a role in functions such as attention, movement, perception, learning and memory. Such synchronisation may also coordinate the actions of the Global Workspace, bringing together information for conscious perception. The theory here is that nerve cells that fire together must be doing something useful together. There are multiple sets of oscillations of neuronal firing across the brain, which can be working simultaneously across these brain regions. Some of these sets of oscillations, for example between slow and fast rhythms of activity, may also be coupled together to bring added sophistication in communication and coordination between neural circuits. In this way, coordinating the activation of neurons across brain regions may help to marshal neural resources to imagine future states, which may be important for other behaviours that propel us towards a range of goals.

This fits well with the idea that the brain creates projections, sampling through memory when looking back in time, and tapping into multiple resources to form predicted projections about the future. We live in the moment, and recall the immediate past, but can also form accounts of concepts and outcomes that impact our near and distant future. This indicates that the human brain has the capacity to escape the prison of immediate sensation. This can help us with tasks that go from the mundane, remembering what we did yesterday, where we left our keys and how much fuel was in the car, through to considering what to prepare for dinner or where to go tomorrow on a country drive; through to more creative exercises, like planning a change in furniture placement in the living room, or contemplating what we might do on the potting wheel at next week's class. This is a generative capacity of the brain, based on experience as well as our ability to manipulate imagery, relationships and concepts internally before or during a related behavioural engagement. Depending on the precise form of imagery, imaging studies have indicated a variety of different brain areas are likely to be involved, including the visual cortex, hippocampus and frontal cortical areas. It is interesting that both primary and secondary visual areas of the brain appear to be involved in mental imaging, supporting the idea that these centres have an important role in producing visual percepts and are not just way-stations along the path to processing visual information for 'higher' brain centres to construct a visual experience.

Mind wandering

Another set of neural pathways, known as the default mode network, involving interconnections between the prefrontal cortex, posterior cingulate cortex, precuneus and angular gyrus, may also have a role in mental imagery and imagination-like brain events. The default mode network becomes active when we are not particularly involved in task-related mental work. In this way, the network is most active when we are not doing much at all – when our minds may be wandering or daydreaming. Conversely, it becomes much less active when we are engaged in highly focused tasks or when our attention is engaged. It does seem to have a role in the flitting mental events that revolve around remembering and imagining, internal narratives, our relationships to others, and autobiographical dimensions such as recent experiences, reflections and sense of wellbeing. Overall, this fits with a rolling internal narrative and introspection that fills up our mental space between more directed mental tasks.

The concept of default mode network for resting brain work is not without its critics. For example, many of the imaging techniques used to identify the network rely on very small changes in blood flow, and it may be that these anatomical regions share blood flow dynamics rather than being meaningfully interconnected. This doesn't really explain how the default mode network shows up on other kinds of scan, such as Positron Emission Tomography (PET), that don't rely on blood flow. Another criticism is that this network is active when scientists don't provide a task, say when the person is staring at a blank screen, so we don't really know which behavioural features are precisely linked to its activity. Others have argued that this network isn't as passive as claimed, and likely has a role in generating autobiographic memories, planning ahead, and cognitive processes related to social interaction. In this regard, the default mode network may help us to sort and understand the emotions and feelings of others, known as 'theory of mind'.

Projecting the past and future

Overall, we can see that there are similarities in how the brain is engaged in encoding, storing and retrieving memories with neural processes involved in considering the possibilities of the future. Both represent the brain projecting itself into past and future states, building cohesive and often multimodal accounts of its physical and social position relative to the external environment. This helps to stabilise and regularise our interactions with the physical and social spaces we encounter. It is clear that this can be done in a focused manner when we attend to a specific memory or future concept, but can also be running at a quieter level in terms of autobiographical, relationship and navigational updates on a constant level when we are not particularly engaged in a specific mental task.

The brain, locked away in the skull, does not just receive information through various sensory modalities, but is active in seeking to be 'present' in the external space - i.e. the machinery of the brain projects a representation of itself into spaces beyond its physical constraints. In this regard, the brain is active in creating a running percept based on information it receives and predictions it makes. The sensory systems that process information do not lead directly to a percept; rather, the brain is constantly creating a percept that efficiently draws on hierarchical sensory systems, seeking to place the construct in the external spaces outside the skull.

According to Projection Space Theory, substantial regions of the brain are not involved in passively processing sensory information for higher levels of integration but are engaged in creating the perception based on what inputs it can sample. The brain is running a mapped simulation of the external world that projects it into physical and social environments.

Bibliography

Dickerson BC, Eichenbaum H (2010) The episodic memory system: neurocircuitry and disorders. *Neuropsychopharmacology Reviews* 35: 86-104.

Vollan AZ, Gardner RJ, Moser MB, Moser EI (2025) Left-right alternating theta sweeps in entorhinal-hippocampal maps of spec. *Nature* 639: 995-1005.

Chapter 7 Projecting to ourselves and each other through language

From its dark, moist skull cave, the brain forms projections of the outside world, and our behaviour is derived and planned based on the sampling of information from the domain outside the skull. It can intentionally coordinate muscles throughout the body, but the brain has also learned to manipulate other objects to reach its goals, which sometimes involve an expansion of the projection space.

One key way for the human brain to communicate with the other human brains is through the use of language. Non-verbal cues and signals received through vision or a touch do help support communication, but the structured use of sounds in particular combinations, or their representation through writing, has become a highly useful tool both for social interactions and the sharing of information. We use language to project out into the physical space beyond the brain in the skull to engage verbally with other humans and via language-based records and accounts of the world.

The regions of the brain that are important for language were initially elucidated by studying the impacts of localised damage to the organ, particularly through strokes. Injury to these brain areas, many of which are named after the researchers who discovered them, produced neurological deficits that affected specific language features. For example, Broca's area in the left inferior frontal cortex is involved in speech production, whereas Wernicke's area in the left temporal lobe, adjacent to the auditory cortex, appears to have a role in the understanding of speech. A substantial bundle of axons known as the arcuate fasciculus connects these two areas, and likely helps translate our understanding of language to centres involved in providing a motor response, for example, a spoken response.

It is also clear that language is extraordinarily complex, and likely makes use of information processing and information generating circuits of the brain. We need to consider that many forms of language are possible, and humans can flip between languages if so trained, which could include integrating non-verbal cues and other forms of sensory and internal state processing, and may not even rely on audible mechanisms, as in the case of sign language. There is also our own internal monologue, the very personal and inaccessible self narrative that runs in either a focused or unfocused, responsive fashion. Finally, there is language communication in other forms, such as in reading where visual information is transferred into an internal narrative.

Language processing involves analysis of a sequence of elements that can be built up into more complex features based on syntactical rules. This involves systems responsible for 'semantic' processing – the meaning ascribed to the various elements. There may be different anatomic pathways responsible for semantic processing and the sequence-based processing of elements into the flow of language. There appear to be pathways in prelinguistic children and other primate species that correspond to simple rule-based sequences, in addition to neural pathways where the hierarchical processing that underlies more sophisticated human language capacity may lie. The hierarchical pathway must maintain a high degree of plasticity over many years, as individuals build their skills in language processing. One of the many reasons we place our children in schools for 10-12 years is to build this capacity.

Language processing plasticity may not end with early life schooling, as it is clearly possible to develop these skills in later life, for example to learn additional languages. At the University of Tasmania, researchers including this author have undertaken a long-term study of older adults (>50 years of age), the Tasmanian Healthy Brain Project – considered in greater detail later in the book – which has shown that engagement in further tertiary level education leads to improvements in language processing capacity and possibly other cognitive domains. Hence, age is not a barrier to continue to build this brain ability.

Combining semantic knowledge with rules-based hierarchical processing of elements into more complex forms and sequences may also be important for other brain functions, such as for music or mathematics. These other functions rely on other neural areas and pathways, but it is interesting that Broca's area in the right hemisphere may also be involved in music processing.

Most of us experience an internal monologue or 'narrative', which may involve activation of brain areas in the inferior frontal lobe. The default mode network may also have a role in the stream of relatively aimless self-talk that some of us experience between being focused on specific cognitive tasks. The internal narrative may help us to mull over potential actions or reflect on events, support working memory to maintain pieces of information 'front of mind', and align with tasks such as reading. 'Self-talk' may also be important for our mental health and wellbeing, either in a positive or negative fashion. Contemporary psychological practice often involves rehearsal of positive internal narratives, with the goal of these becoming more entrained over time.

However, not everyone has an internal monologue, and this may also be associated with difficulties in generating internal mental visual images.

The Mind's Eye is missing!

Some people have difficulty with internal visual imaging, known as 'aphantasia'. Many aphantastics also have difficulty with internal mental representations of other sensory modalities, including auditory imagery such as the internal narrative. For most people, aphantasia does not appear to be linked to any particular brain insult or damage, although it can arise following various forms of brain injury. Aphantastics may have difficulty in producing vivid mental images, but they can frequently describe what such images should look like if prompted. This shows that they understand what may comprise an image, indicating preservation of semantic processing and memory. Aphantasia does not appear to be associated with major deficits in brain function, but is sometimes associated with poorer autobiographical memory and face recognition. It has also been linked with higher levels of autistic traits, particularly in imagination and social skills. Otherwise, it seems likely that many people with degrees of aphantasia may not even be aware that they have less ability in forming mental images relative to the wider population. Other forms of internal mental imagery, such as occurs during dreaming, may remain intact.

Similarly, there are people with limited or no internal auditory imagery ('anauralia'), sometimes with no internal narrative ('anendophasia'), though research has been relatively limited. People with these conditions live 'normal' lives, however, without any particular major deficits in functions that we think may require an inner voice. They often report that they will express themselves out loud with spoken language if they need to 'hear' an internal reasoning process or when they read.

These conditions show that there are likely many ways in which the brain can 'think', process information and choose actions, that do not require internal imagery or a narrative. Conversely, internal representations of 'thought' may represent a veneer of conscious experience that may not be necessary for the human brain to perform a variety of functions.

What is the neurobiological basis for the generation of ideas and concepts that are then translated so seamlessly into language? Where does the pre-production part of communication lie in the brain? Research suggests that there may be a pre-linguistic stage in terms of language production, possibly involving centres in the anterior temporal lobe, leading to activity in the prefrontal cortex where the phonetics and planning of speech occur. There have been remarkable recent studies involving the use of recording arrays applied to the human brain while the subject is awake. These have led to the identification of neurons, for example in the prefrontal cortex, that encode 'phonemes' – the smallest unit of spoken language – as well as syllables. They show that there may be a sequence in which these elements are constructed before a word is spoken. Furthermore, the research reveals very little overlap in activity of cells involved in the planning and perception of speech, indicating that these processes are complementary within the same language centre.

It may be that language flows from previous prompts or stimuli – that it is a response to a preceding sequence of events – but that doesn't really account for intentionality or originality in forming an idea requiring a language component. While we may on occasion go from internal narration to speech production, it would seem likely that the generation of language mostly sits in neural networks to which we don't have conscious access. While we could use an internal narrative to ideate or rehearse external communication, the first time we have a conscious appreciation of the things we say may be after we have said them. This may feel a bit unsettling but aligns with other brain functions where we comfortably rely on non-conscious neural processing to occur less than a second before they enter consciousness. Spoken language may be another form of projection of the brain, where ideas, concerns, or information enters into a conscious space, potentially so that we may communicate with other brains equally marooned in their skull boxes.

Language may have provided a distinct evolutionary advantage for such human-to-human communication, with written forms arising to help provide historic accounts of events or principles to communicate more widely and/or as records for future generations. Beyond auditory processing for speech or visual processing for text, the flow of conscious experience that is associated with our ability to communicate with each other represents the projections of brains into the space between our brains.

Bibilography

Khanna AR, Muñoz W, Kim YJ, Kfir Y, Paulk AC, Jamali M, Cai J, Mustroph ML, Caprara I, Hardstone R, Mejdell M, Meszéna D, Zuckerman A, Schweitzer J, Cash S, Williams ZM (2024) Single-neuronal elements of speech production in humans. *Nature 626*: 603–610.

Littlejohn KT, Cho CJ, Liu JR, Silva AB, Yu B, Anderson VR, Kurtz-Miott CM, Brosler S, Kashyap AP, Hallinan IP, Shah A, Tu-Chan A, Ganguly K, Moses DA, Chang EF, Anumanchipalli GK (2025) A streaming brain-to-voice neuroprosthesis to restore naturalistic communication. *Nature Neuroscience* 28: 902-912.

Chapter 8 Projecting through tools and technology

In the previous chapter I argued that language projects human minds into the space between us. Another marvellous feature of the human brain is our capacity to work through tools to interact and manipulate our environments. It is important to recognise that non-human animals have also developed the capacity to work with tools, but it has been human ability in this area that has resulted in entirely new capacities and technologies that have advanced human civilisation – not always to the ultimate benefit of the human species, other species, or the planet. Unlike tool use in other animals, the application of technology in humans, and its improvement over time, far exceeds the imperative to feed and breed. Rather, tools are used to extend the capacity of humans to manipulate their environment in ever expanding ways. Some anthropologists have argued that tool use may have evolved alongside language, because language would serve to help pass on tool-making skills, and that neural systems for tool use by the hands may have migrated to the use of mouth muscles for speech. There is evidence from neuroimaging studies that, for example, that brain regions for hand and mouth movements may be closely inter-related. Notice next time you are threading a needle - you may also be slightly poking your tongue out! A converse view is that perhaps human progression through tool use drove greater levels of interaction between people and groups, leading to advantages that arose from better communication via language.

Tool use makes use of a complex interplay of a variety of neural systems for sensory feed-back, predictions of consequences of certain actions, and the planning and sequencing of motor functions. Much of what is known about the neurobiology of tool use comes from studies in non-human primates using simple tools to manipulate their environment, such as using a rake to extend their reach to obtain a food reward. These studies showed that, after training, receptive fields of neurons that represented the digits could expand to include the rake tool. In similar studies involving grasping for food using pliers, neurons in the premotor cortex that were normally activated for hand grasping were mobilised for the use of the pliers. This indicates that the brain is adept at developing a neural representation of external objects we meaningfully interact with – that we can extend our brain activity to encompass a desired manipulation. This aligns with the concept that neural structures can model and adapt to position or project ourselves out into the world.

The brain's understanding of causality – if I do action A with item B, this will result in outcome C – probably underlies its success with tool use. This aligns well with strong interest in the predictive capabilities of human brain circuits, imagining and pre-empting future states based on current state actions.

Brain activation studies suggest that a region of the cerebral cortex known as the anterior supramarginal gyrus, located in the parietal lobe at its juncture with the frontal and temporal lobes, plays an important role in tool use. This is one of those brain regions that receives a range of sensory information – a polymodal association area – including input from auditory, somatosensory and visual centres. It shows increased activity also in the planning stage of tool use, and lesions in this area can result in an inability to make certain movements, a condition known as apraxia. Tool use must also involve a range of other brain areas, particularly in motor planning in the frontal cortex, as well as in relation to the complexity and ordering of the task and recruitment of other required brain functions.

Neural maps for tools

When we learn to use a particular tool, our brain activity may map to the tool in the salient brain areas in much the same way it aligns to body structures such as limbs, hands and digits. Our capacity to extend neural maps and representations through the tools we interact with may explain the phenomenology of projection of sensory states. In this regard, we 'feel' the tyres and position of the car we are driving, the tip of our pen on a piece of paper, the glide of an iron over clothes, the stitching together of fabric through a sewing machine and so on. This is the kind of capacity that also underlies how various haptic technologies can provide a simulation of a sensory experience. Vibrations at certain frequencies in gaming devices, for example, can stimulate mechanoreceptors to give a sense of contact. The laptop I am using for writing this book has a trackpad which does not move but provides haptic feed-back to create the sensation of 'clicking'. Haptic feed-back technology is also increasingly used for virtual reality applications, including for operating robotic tools over a distance.

All this shows that when a brain develops a picture of the world outside the skull it is not fussy about the form of sensory sampling it uses to help project an account of its physical reality *extra cranium*. We have this capacity to project through a range of forms of sensory stimulation, with the process of projection causing a conscious percept.

This may explain how we also interact so deeply with other forms of technology. An example of this could be reading a book. Reading usually involves immersion into the text, using our understanding of word meaning, syntax and grammar to create a flowing internal mental representation of the text such that our brain works to produce mental images and sometimes other sensory representations of the world from the narrative, anchored in memory from previously read book sections. We become embedded in the narrative such that it feels like an experienced account, and not just word sequences. This phenomenon is well known in writing circles as 'narrative transportation' when we feel that we are present in another world.

This is a good example of how the brain may project itself through technology to create a conscious experience. Similar experiences may arise from viewing a movie: rather than breaking the images and sound into its constituent elements, our brains seamlessly transport us into the story (if it is a good and engaging film). We don't need to be trained to do this, it is an easy process as our brains are adept at extending themselves into the external environment, with the act of extension creating the conscious flow.

This is an extraordinary ability that we take for granted in our daily lives, and the synergistic relationship between human brains and technology has likely not yet reached its limits. We appear to be at an early phase of increasingly direct integration with machines, with a variety of innovations that link technology with our neural machinery. On the lower technical dimension of this is the development of effective prostheses to replace limbs, which allow for substantial brain adaption both in sensing and operation.

Transported into other worlds

At the other end of the scale, there is escalating sophistication in smart devices, such as phones, tablets, glasses and watches, that involve increasing degrees of functional integration with our brains. Clearly, our brains are capable of extending through such devices, not only to generate useful information but also for seamless engagement in conscious experience. Here the technological hack is not just about device hardware, but also in the software that draws our brains in and commands our attention. In this way, social media scrolling effectively entraps our brains into cascading accounts of the exterior world in which we come to feel directly engaged. We are projected through a device into multiple worlds, whether it's kittens at play, celebrity shenanigans, the support or dispute of political views, or breaking news from across the planet. The somewhat manipulative trick at play is that we are not just passive observers, but that our brains make us feel like happy participants craving further stimulation, often in short digestible segments. This is very similar to the concept of narrative transportation where an excellent book or film makes us feel like we are living in another world. A recent systematic review and meta-analysis on the effects of such 'short form video use' has demonstrated a correlation between the amount of exposure to this format and poorer cognition (particularly attention and inhibitory control) and mental health (stress and anxiety). The neural colonisation of these virtual worlds may have a cost to the effectiveness of our brains in other domains.

While such smart technologies bring several opportunities to sample external worlds, there is a risk that addiction to bite-size pieces of immersion will degrade our ability to attend for longer periods of time or have healthy interactions with the external world. This is a 'natural experiment' for humanity on a grand scale, and we will soon be able to determine how years and decades of reliance on these digital prostheses impacts brain capacity and function.

Another major challenge for human integration with technology has been in the development of brain-machine interfaces. A famous and successful example of this is the cochlear implant, which corrects hearing impairment. An external sound receiver processes sounds and transmits them to a receiver under the skin, which then sends signals via fine wiring to an electrode array implanted in the cochlea of the inner ear, resulting in direct stimulation of the auditory nerve. The success of such implants depends on the ability of the brain and its auditory pathways to adapt to this form of signalling, in order to make sense of the artificial stimulation. The cochlear implant is often used as a model to inspire other major technology innovations such as attempts to develop a bionic eye.

While cochlear implants represent impressive technology that has potential to seamlessly integrate with our nervous systems, the use of such approaches is not without criticism and ethical concern. For example, who has access to both the technology and surgical procedures of such bionic approaches? Further, some members of the hearing-impaired community feel that corrective approaches impact their collective identity as people who have effectively adapted to hearing issues: an example of the ableist views of the wider community seeking to correct or medicalise difference.

Another example of rapid growth in brain-machine interface technology is where electrodes, or arrays of electrodes, directly contact the brain, either across its surface or directly into brain tissue. There are many areas in which this approach is being pursued. The goal may be to record and understand neural information from motor areas of the cortex to stimulate peripheral muscles, enabling people with spinal injury to move limbs again. In an extraordinary example of a brain-machine interface restoring function, a recording array applied to the speech sensorimotor cortex combined with machine learning has enabled speech synthesis in people who are unable to vocalise due to severe paralysis following a stroke. The brain data is collected and transformed in 80 millisecond tranches and the output is passed to a voice synthesiser. While in its initial stages of development, the potential applications of this combination of recording and AI-based analysis signals new frontiers in brain-machine prostheses and other collaborations.

Other versions of these brain-machine interfaces involve stimulating very specific areas of the brain to control neural activity to reduce symptoms of conditions such as epilepsy and movement disorders, particularly Parkinson's disease. Brain-machine interfaces have also been trialled to control psychiatric conditions such as depression. Military organisations have taken an interest in this technology, with cyborged insects, rodents and birds potentially playing roles in surveillance and delivery of ordinance. Weaponised and remotely controlled animals may be a feature of future wars. Reversible forms of these brain-machine interface approaches may potentially sharpen and augment the senses and motor responses of soldiers in the field.

Ethical guardrails

Both the physical and software elements underlying brain-machine interfaces are becoming increasingly sophisticated, including miniaturisation and the minimisation of impacts on the brain, mathematical modelling of information transfer and the application of advanced machine learning algorithms. We are not yet in the realm of the *Ghost in the Shell* anime, or jacking into *The Matrix*, but the possibilities of brain-machine interfaces integrating brains with technological tools seem boundless. This does raise the issue of regulation – where are the medical and ethical guard-rails to ensure that bionic augmentation doesn't transgress either actual laws or people's dignity and wellbeing? While approaches such as deep brain stimulation have shown efficacy in controlling motor symptoms in movement disorders such as Parkinson's disease, to what degree are we comfortable with such approaches being used to regulate our moods or other brain attributes? Are we prepared to accept potential 'off-target' psychological changes that may occur as a consequence of our brains being stimulated or controlled? At what point does such technology become more 'cosmetic' in ameliorating traits we might find inconvenient, or where we look to augment our nervous systems as our ableist society has difficulty with accepting people with disabilities?

Beyond disability, how do we feel about augmented human beings who may have technology-enhanced abilities? It isn't too fanciful to predict that some people may prefer a deeper level of integration with, for example, social media, or to enhance their experience in gaming or sex. In the absence of meaningful regulation of such frontier technologies, it is easy to see how technocrats might seek to commodify and exploit our brain's ability to project through such technological tools. As I review some health stats on my Apple watch, and check in on my iPhone, it is hard not to consider that we are already well on our way to such a bionic future where we become more and more projected through, and integrated with, machines, whether or not it involves electrodes in our brains.

Bibilography

Nguyen L, Walters J, Paul S, Ijurco SM, Rainey GE, Parekh N, Blair G, Darrah M (2025) Feeds, Feeling and Focus: A systematic review and meta-analysis examining the cognitive and mental health correlates of short form video use. *Psychological Bulletin* 151: 1125-1146.

Vickers J, Gilbert F (2013) Brain danger. *MJA Insight* 2: 1

Chapter 9 Artificial intelligence and the brain

Artificial intelligence or AI poses a range of challenges for human brains, and the jury is still out as to whether it benefits or harms us, or some combination of both. The miracle of the vast information source that is the internet has already made it much easier for our brains to access knowledge and facts. AI turbocharges this library of information by providing an automated analysis of all of this data. Conceptually, AI has been around for several decades, but the explosion of learning models that use generative AI or 'genAI' to create original content – including text, images or animations – has already impacted many areas of human endeavour.

There are a variety of different AI processes, but let's focus on the underlying principles of genAI. This model relies on algorithms to tie together all of the elements and parameters of large data sets. In this process, often referred to as deep learning, the system can build new outputs based on the relationships between these elements, allowing it to construct sentences and paragraphs, outputs that appear to make sense relative to a specific prompt. The underlying systems for AI are often referred to as large language models, and they involve computing-intensive processes with specialised hardware to establish the connections between all of the elements in large data sets. Outputs from these large language models can be assessed for their accuracy and performance, leading to fine-tuning or 'training' of subsequent outputs. Overall, genAI produces a facsimile of intelligence or understanding based on its ability to tie together underlying data relationships. The end-user may access this database by providing prompts to derive particular outputs. These outputs are not necessarily static as they can be refined by adding more iterative prompts to 'fine tune' the required output.

The simplest way of accessing AI has been through the use of 'chatbots' or 'AI Assistants' that largely rely on text enquiries to produce outcomes, very much under the control of human operators. The field has since evolved into interfaces referred to as 'agents', which can undertake complicated goal-orientated analyses in a more sequenced and autonomous way to produce outputs that may involve integration with other systems. The outcome of this approach often relies on the sophistication of the language prompts provided to the agent, as well as subsequent iterative improvement by to-and-fro communication with the agent. The autonomous and learning nature of 'agentic' AI, where AI agents generate their own feed-back loops and integrate with other systems, provides even greater opportunity to automate a range of processes, potentially revolutionising areas such as manufacturing and the provision of services. The scale of AI's impact may be as significant as that of the Industrial Revolution, with replacement of human job roles but also new classes of workers who are able to make use of the symbiotic relationships with these programs.

A facsimile of learning

Developments in AI also expose some of the deleterious aspects of the modern educational system. In universities, for example, education is largely oriented around the assessment of assignments and essays produced through word processing programs and submitted online. These programs already assisted students by correcting spelling and grammar, but AI has brought a significant new challenge to this way of assessing learning. With the right prompts, AI engines can produce assignments and essays that look convincing in terms of their analysis of data sources and the production of 'insights' in a facsimile of understanding and critical analysis. In this way, a student can produce and submit an educational product that has involved no reading of books or research papers, no analysis of underlying themes or concepts, and no integration of such knowledge into novel observations.

The modern corporate university that treats its students as customers seeking credentialing may be tempted to engage substantially in AI to reduce the costs of providing education. This could involve using AI to generate content for courses, as well as the use of AI in assessing students' work. Perversely, this may lead to students producing AI augmented assessment items that are then critiqued by AI, with both students and teachers sitting on the sidelines of the process, maybe providing some guidance or guardrails.

What is the answer to the pernicious impact of AI in education? Much has been proposed about adapting assessment practices to minimise possible AI-related augmentation, such as a return to more invigilated and hand-written forms of assessment, or the revival of one-to-one relationships between teachers and students to allow direct enquiry into how well they have learned and understood their courses. The growth of on-line learning may need to be revisited and innovations adopted to ensure that students are producing authentic assignments that represent their learning.

It may be that AI will drive a transformation in both how educators teach and how we assess what students learn. While we may yet see a return to more personal forms of education, such as 'Socratic' approaches that involve a dialogue between teacher and student, it is not certain how practical this may be in educational systems that are being optimised to 'process' as many students as efficiently as possible. Alternatively, it may be possible to adapt AI engines to encourage critical thinking and learning, so that the AI enables learning rather than replacing it.

AI is here to stay and its impact on our lives will only increase in future, and so it makes sense to ensure that education includes development of skills to ensure effective and meaningful engagement with its algorithms. What may be required is a new paradigm that promotes the creative use of AI in education.

A seamless symbiosis

Given that AI mimics human intelligence, it will be easy for our brains to make use of these new tools to project themselves into the world in a seamless symbiosis between human and machine. For example, it may not be necessary for a future accountant to understand all taxation codes provided they can ensure that the data of clients is inputted appropriately into an income tax assessment. Agentic AI can then move the process along further by 'consulting' with Tax Office AI programs to produce a final outcome, depositing returns automatically into bank accounts. In our hospitals, all past medical history as well as 'live' diagnostic data might be fed into AI programs that can assist in the diagnosis of conditions and suggest treatments. Robot surgeons, trained on large data sets, may then be able to undertake the appropriate interventions, potentially under the guidance of human specialists remote to the hospital.

Our AI future may be more dystopian on the battlefield. Drone technology is already having a major impact on armed conflicts in many regions of the world. As this moves further into the 'agentic' space where drones may operate much more autonomously, this raises the spectre of human-less, technologically-driven war games where the outcomes will be determined by whoever has the more sophisticated and adaptive AI-based approach, as well as the most effective factories turning out drone warriors.

The opportunities and challenges posed to our human brains by the AI revolution are hard to predict. We are already connected to AI though our computers, phones and other smart devices. AI-generated online content is also becoming more pernicious, often making it difficult to determine what is 'real' and what is not. We have seen how good the brain is at projecting through the use of tools, which includes digitally-based tools that work so seamlessly with our neural machinery that they can ensnare our attention. It is evident that AI has great potential to augment our engagement with the machines, as we increasingly come to rely on their processing power. While this may improve human outcomes in a range of areas, there is a real risk that reliance on this digital form of intelligence may be deleterious to brain function. 'Cognitive offloading' of tasks to our machine partners may result in our losing the capacity to digest and process information. Preliminary studies have already indicated the negative effect of AI use on human learning and adaptability. Short term benefits may follow, but given the plasticity of the human brain, it may be that outsourcing brain work to this form of technology leads to a 'dumbing down' of the neural networks that currently provide a range of human functions. Will this symbiotic relationship be a form of asymmetric development, where the AI programs become more effective at replacing human functions at the price of human brains becoming less sophisticated? The inherent adaptive biology of the human brain would suggest that there is a significant risk involved in outsourcing cognitive and intellectual tasks.

Conversely, a study from the Boston Consulting Group showed that workers engaged with AI tools may also run the risk of cognitive overload as they seek to deal with the torrent of information generated by AI agents. They coined the term 'AI Brain Fry' to describe the mental fatigue that can result from excessive use of AI tools beyond the capacity of the human brain to accommodate.

Whole brain emulation

Another broader concern is that AI development will lead to programs and machines that have conscious experience. This would seem to be an unlikely outcome. There is no question that AI is becoming more powerful and sophisticated, but this still represents only a simulacrum of human brain capabilities. For as long as we lack a precise understanding of the neurobiology of human consciousness, the capacity to reproduce this in programming or the architecture of chips and machines will also be lacking. A goal of many of the major players in AI is to build an artificial general intelligence that is highly autonomous and adaptable, and can mirror and surpass a range of human-like intellectual and cognitive abilities. Proponents of 'whole brain emulation' suggest that a brain simulation, based on the neurobiology of the human brain, may be a route to artificial general intelligence, but there is still a mismatch in the architecture of computers relative to our limited understanding of human brain operations at the nerve cell level and the 'connectome' – their multiple, highly complex connections.

Progress in this area will also be limited if we focus on the information processing aspects of human brain function, rather than its information producing capacity, as would be necessary for a simulation of the projective nature of the brain. In this regard, an understanding of the modular unit of cortical function, particularly in terms of recursive circuits, may provide a technological guide to future processing architecture. However, given the existential threat inherent in a value-less artificial general intelligence, our gaps in knowledge of how human brains work may be a fortuitous brake on the development of autonomous machines that could be indifferent to their human creators.

In Greek mythology, Zeus did have his issues with Prometheus and Pandora when they unleashed fire and a variety of ills and evil, respectively, upon humanity. The lesson here though is that when these 'gifts' were out, they were out, and it was up to humans to decide how they would impact their lives and development. There is no question that AI is also out, and the extent to which it changes the course of history will likely be revealed very soon, and will depend on whether meaningful guardrails are put in place. Clearly, this will bring the symbiotic relationship between the world of machines and human brains closer and closer, for both good and ill. If the machines can build a capacity to project themselves into the external world in a similar fashion to a human brain, then the possibility of conscious non-biological entities may move out of the realms of science fiction into reality.

Bibliography

Bedard J, Kropp M, Hsu M, Karaman OT, Hawes J, Kellerman GR (2026) When using AI leads to "Brain Fry". Harvard Business Review https://hbr.org/2026/03/when-using-ai-leads-to-brain-fry

Kosmyna N, Hauptmann E, Yuan YT, Situ J, Liao X-H, Beresnitzky AV, Braunsten I, Maes P (2025) Your brain on ChatGPT: Accumulation of cognitive debt using an AI assistant for essay writing task. MIT Media Lab. Available on *arXiv:2506.08872*/ https://www.brainonllm.com

Chapter 10 Building better brains

One of the great challenges facing humanity is the substantial increase in the prevalence of dementia across our communities. Dementia is an umbrella term that is used to describe the changes in brain function associated with dozens of different neurological conditions, the most common being Alzheimer's disease. It is tragic that this condition slowly robs people of many of the unique human capabilities that come with our sophisticated brains. With Alzheimer's disease, we know that specific groups of neurons are preferentially affected. In the cerebral cortex, this selective vulnerability includes pyramidal cells, and their gradual loss leads to a progressive loss of synaptic connections, resulting in a slow disconnection between brain regions. While a person with Alzheimer's disease still have 70-80% of their original neurons in the cerebral cortex, the selective loss of connectivity in terms of both local and long-distance projections is sufficient to degrade brain function.

Most people associate Alzheimer's disease with problems in memory, particularly forming new memories. However, the disease process is progressive and undermines a range of brain functions, such as planning, orientation to place and time, spatial navigation, personality and behaviour, and communication. By the end stages of Alzheimer's disease, a person with the condition may show very little connection with their external world, lack any substantial communication, have very limited mobility and require assistance with functions such as eating, dressing and hygiene. Nonetheless, there may be limited periods of lucidity and evidence of being conscious of their surroundings.

It has been argued that Alzheimer's disease is a condition that specifically erodes a person's conscious capacities, including recognition of conscious processes in others as well as within themselves. This may potentially explain difficulties such as awareness of internal states, as well as communication. In the context of Global Workspace Theory, one of the leading theories of consciousness, Alzheimer's disease pathology degrades the corticocortical connectivity that 'ignites' conscious experience. Conceptualising Alzheimer's disease as a disorder, or altered form, of consciousness, may provide a prism for understanding many of the deficits that arise. It may provide a guide for approaches to care and communication, as well as potential and possible rehabilitative strategies to enhance conscious capacity.

Alzheimer's disease is not a condition that is caused by ageing, but it is associated with ageing. In other words, it does not result from general wear and tear of the brain over many decades but is more linked to pathological brain changes that are associated with ageing. One particular research focus is the build-up of sticky fibrillar proteins (known as amyloid) in clumps or 'plaques' between nerve cells. These amyloid plaques are largely comprised of altered versions of a protein that is normally present in the brain, referred to as the beta-amyloid protein. Plaques build in numbers over time in step and cause damage to nerve cells, particularly to their process such as axons. There may be many years between the formation of plaques between nerve cells and the point at which they cause sufficient damage to impair function . Given that it is likely there is a high degree of redundancy in connectivity in the brain, there can be some compensation, but then a threshold is reached beyond which the brain can no longer compensate.

There are a number of different ideas about how beta-amyloid may cause damage to nerve cells – enough to write a book solely on the topic. There has also been a degree of push-back on the primary role of beta-amyloid as the cause of Alzheimer's disease. Some researchers have argued that alterations in beta-amyloid and plaques are more likely to be a consequence rather than cause of the disease process. There is a view that internal changes within neurons, particularly involving a protein called 'tau' may be more important in the pathological sequence leading to nerve cell degeneration and dementia.

The balance of evidence from a number of different approaches, from studying human brains through to familial or genetic forms of the condition, indicates that beta-amyloid is an early feature of the disease. In this regard, it may be more of a trigger for neurodegeneration rather than a driver of neuron death. Research at the University of Tasmania has shown that it may be particular kinds of plaques that damage nerve cells, that most of this involves injury to axons and synapses, and that this triggers an abnormal regenerative response that leads over time to alterations throughout a subset of neurons, a loss of connectivity and their eventual degeneration. However, this is just one theory, and there are a multitude of ideas that relate to the association of beta-amyloid deposits in the brain with the progressive and degenerative brain changes that underlie dementia.

Recently, we have seen the advent of 'disease-modifying therapies' based on monoclonal antibodies directed against beta-amyloid, and where these have worked in clinical trials it is in the modest slowing down of disease progression, not halting it. This is a welcome outcome, but it is also notable that these new drug options come with the a significant risk of adverse side effects, such as brain swelling or bleeds. It is hoped that the monoclonal antibody approaches may become more effective with further development, with reduced risk of potentially serious side effects.

A focus on prevention

In the meantime, there has been increased focus on potentially 'modifiable' risk factors for dementia. Systematic reviews and meta-analyses conducted for a Lancet Commission report in 2024 suggest, that up to 45% of dementia cases world-wide could be avoided if 14 of the risk factors were eliminated from the population, with each of these factors having particular importance at specific stages of life. These include low educational attainment in early life; hearing loss, high LDL cholesterol, depression, traumatic brain injury, physical inactivity, diabetes, smoking, hypertension, obesity and excessive alcohol intake in midlife; and social isolation, air pollution and vision loss in later life. The particular importance of these risk factors is likely to vary across cultures and populations throughout the world, but this does provide some hope that by targeting these risk factors for improvement, we may see a decrease in the incidence of dementia as we age.

As with a range of health conditions that are linked to lifestyle, there is often a focus on the individual and what they can do to identify and reduce a specific risk factor. However, changing health behaviours related to these factors can be hard! In the case of factors that relate to sensory systems, such as hearing or vision impairment, the response may involve testing and the provision of an aid or procedure. Other risk factors – including obesity, high blood pressure or LDL cholesterol – may be more difficult to address. Perhaps the place to start here, and with many other factors, is the identification of whether an individual has these risks. It could be advantageous if primary health doctors were able to introduce screening for their patients, ideally to commence in midlife, where many of these risk factors may start to have an impact on brain health.

It is also important to recognise that these risk factors are not just about making healthy lifestyle choices but are also closely linked to environmental and social policy. Broadly, living in disadvantaged circumstances with limited economic resources is associated with many of these risk factors. For instance, living in areas with high air pollution may be difficult to avoid where employment or housing opportunities are concentrated in these places. Risk factors are also likely to be inter-related. For example, smoking and physical inactivity are associated with general poor health, and may factor into rates of obesity and risk of high blood pressure. Hearing aids and surgical correction of vision loss may be beyond an individual's budget, and both factors may contribute to social isolation.

Overall, these risk factors are likely to be at higher rates in communities with lower levels of education and general health literacy. Conversely, when we have people sign up for research studies related to these risk factors, they tend to come from well-educated and higher socio-economic communities, which in general have greater access to health resources and lower rates of many of these risk factors. This means that reducing dementia risk in the wider population will involve not only building awareness of these risk factors and their impact on the brain, but addressing specific community and policy settings.

There is evidence that through a range of policies and initiatives, we may already be seeing a reduction of dementia prevalence at particular ages. This has not been due to any intentional policies regarding brain health, but may follow broader changes in rules, policies, and health awareness. For example, this may include increases in early life education due to policies around requirements for formal education, reductions of air pollution, more screening for high blood pressure, and better access to procedures such as cataract correction.

New approaches to address health factors linked to dementia risk have also been investigated in studies across the world. Many of these involve delivering structured 'multidomain' approaches to risk that provide targeted advice and programs. These have shown minor benefits for cognitive function, but it is not clear how such intense programs could be delivered efficiently on a sufficient scale to make a difference, particularly with disadvantaged and hard-to-reach community segments. A variation on this approach is to offer programs targeted at specific at-risk populations. This may be undertaken by encouraging people who may be at risk of dementia due to a family history or genetic predisposition, for example, to engage with a risk reduction program.

Traffic lights for risk

In Tasmania, which has a relatively 'old' population, the Wicking Dementia Research and Education Centre where I have worked has undertaken a broader approach to encourage older adults to attend to modifiable risk factors. In addition to educational sessions for the general public across Tasmania, our Island Study Linking Ageing and Neurodegenerative Disease, aka

'ISLAND', has a structured online support program consisting of a dementia risk profile tool that identifies and provides personalised advice on any risk factors participants may have, as well as access to the Preventing Dementia Massive Open Online Course (MOOC). The dementia risk profile tool brings to the forefront the main areas of risk for each ISLAND participant, and is repeated yearly so that participants can track their progress over time. This tool provides a traffic-light result for risk factors – green for doing OK, red for being in the risky zone, and orange to indicate somewhere in between, probably on the way to turning red over time. The output from this tool can also be shared with health providers, if the participant should so wish, so they can potentially assist in understanding and tackling any areas of risk.

The MOOC provides an understanding of epidemiology and risk, and then presents the main risk factors for dementia that are supported by the best evidence. It also identifies a range of factors that may erroneously pop up as risks for dementia either as a result of misunderstanding or misinformation. The MOOC has been freely available across Australia and the globe since 2016, has reached hundreds of thousands of participants, and our published work indicates that exposure to the course can stimulate risk-reducing actions.

This ISLAND approach has been shown to result in a reduction of risk factor prevalence in participants who follow the program over time and has been particularly efficacious for participants with multiple risk factors. The personalised

Dementia Risk Profile tool has a potent impact on risk behaviours and can work synergistically with exposure to the Preventing Dementia MOOC for risk factor reduction. Substudies that have arisen from the ISLAND cohort include: new diagnostic approaches involving AI-based assessment of movement captured by video; the pathways of dissemination of dementia risk-related knowledge in the community; the impact of natural disasters on brain health and dementia risk; and the use of blood-based biomarkers for early screening in the population.

ISLAND demonstrates that risk reduction is possible and can be in the hands of people who are motivated to take action and to develop their literacy in this area of health. It has been interesting to see which factors have been more amenable to change. Factors such as obesity are clearly hard to shift, but participants showed greater capacity in engaging in cognitive stimulating activities, social engagement, alcohol reduction and physical activity. One of the most striking findings to date has been the reduction in the proportion participants showing multiple risk factors, from approximately 10% down to 3%. While we are unlikely to be able to eliminate all of the dementia risk factors from our population, it is seemingly possible to make a meaningful dent in them. It is always important to remember that ISLAND involves a volunteer sample, who are likely to be more motivated in this area of health than the wider population, have good language proficiency, and have greater access to resources such as the internet. The tricky goal here is how to frame messages and promote motivation around dementia risk to people with fewer resources, limited digital access or literacy, or for whom English is not their primary language.

Can we build cognitive reserve?

A different study from the Wicking Centre points to another avenue to promote healthy brain plasticity and to build resilience to conditions such as dementia. The Tasmanian Healthy Brain Project focused on the potential for education for older adults to bolster their cognitive reserve. Education can be considered a form of 'applied neuroscience' to help build our brain capacity. We are all used to the cultural norm of placing our children in a program of education stretching from a young age into adolescence and for many as young adults. In most countries, there are 10 expected years of education, and sometimes as many as 12 or 13 regulated years. For young people who complete these initial school years, there is then the opportunity for further years in education or training, including vocational training or colleges/university. A substantial amount of population data from across the world demonstrates that the more years of education an individual has earlier in life, the less chance they have of developing dementia at advanced ages. A different way of looking at this may be that low educational attainment is a particular risk factor for dementia. Lower levels of this are more prevalent in some low- and middle-income countries; among women who have historically experienced less access to education; and within particular population cohorts where opportunities for extended formal education were more limited.

There is evidence that the educational effect on dementia risk is higher for people with a genetic predisposition for dementia, for example, if they have one or two copies of the epsilon 4 version of the apolipoprotein E gene. There is also the argument that the same genetic background may affect both educational attainment and dementia risk. In other words, the education-dementia risk link may relate to a common genetic predisposition, not necessarily to the benefits of education on brain health. A fascinating study on this education-dementia link was led by Professor Carol Brayne from the University of Cambridge, UK. Professor Brayne worked with colleagues through England and Europe to bring together three longitudinal studies of brain ageing which involved the collection of brains from study volunteers after death. This large, combined study confirmed that greater years of education in early life were associated with lower risk of dementia at death. However, when the researchers looked at the most significant Alzheimer's disease-related brain changes, they found that education did appear to have a protective effect. At a relatively equal level of pathology burden, there was a greater chance of dementia being present with low educational attainment as compared to relatively higher educational attainment.

This means that if you have more education, you can resist the impact of disease burden on the brain, regardless of your genetic predisposition to the disease. Consistent with the concept of brain or cognitive reserve, education likely imparts greater brain resilience, perhaps by ensuring there are multiple or more efficient neural pathways, hence, higher levels of redundancy. This may protect function in the presence of brain injury and disease. The protective capacity of education has also been shown to contribute to resilience to the impacts of other brain conditions, such as stroke, multiple sclerosis, and Parkinson's disease.

Promoting an enhanced cognitive lifestyle

The 10-year longitudinal study that was the Tasmanian Healthy Brain Project involved providing people over the age of 50 with access to university courses (with subsidised fees) of their choice. We found that this did not have a protective effect on some areas of brain function, such as executive function, which involve tasks that rely on planning or the speed of neurological processing. However, we did find a consistent effect over the years on language processing, as well as evidence of a positive influence on verbal memory. This finding was particularly interesting as it relates to the understanding and use of language, an area of crystallised intelligence and a key aspect of cognitive reserve that has been shown to be protective of dementia in other studies. As an aside, this study also showed that ageing (up to the oldest participant who was 79) had no impact on participants' academic performance.

Clearly, the educational experience that individuals have may be quite variable, and competing interests, relative levels of disadvantage, educational/government policies and cost would mean that access to education may be difficult. Other potential pathways could provide many of the same brain benefits. For example, population-based studies have indicated that working longer, in a job that you don't dislike, and in positions involving the use and manipulation of data as well as interactions with colleagues, may provide a protective effect for brain health. Other studies have indicated that a robust 'cognitive lifestyle' that involves reading books, visiting galleries, playing board games and, or learning a new language may be beneficial. These are activities that involve novelty and intellectual challenge, not just repeating familiar activities. The key here may be in the variety of intellectually stimulating activities, not just in doing one over and over. This is supported by a range of neuroscience studies showing that the synaptic connections between neurons form and reform throughout our lifetime. Our own results with experimental animals (mice!) have shown that exposing relatively older animals to enhanced environmental stimulation produces synaptic and plasticity changes in the brain, even in mice that were genetically altered to express some features of Alzheimer's disease.

Taken together, this would indicate that a lifetime of engaging in cognitively stimulating tasks, and perhaps across a variety of activities, may produce the maximum benefit for building resilience of the brain to ageing and a range of harmful diseases and conditions. Combining this with additional social engagement and increased physical activity may be a robust public health recipe to encourage the population to boost brain health. We are probably not quite there with a specific prescription on how much and what types of these activities may be ideal. When it comes to physical activity, however, it is clear that any increased activity is better than being inactive, perhaps with a 'sweet spot' of around 7,000 steps per day in addition to some form or resistance work. Perhaps at this stage the best advice we can give is to increase across all of these areas of activity as much as you can, without causing any stress or harm.

Clearly, there are aspects of this approach that most individuals can engage with, and there are also a variety of policies that governments and public agencies could pursue to support and encourage building our brain health. These could include reducing exposure to air pollutants, incentivising physical activity and social connectivity, and providing opportunities for cognitive stimulation. At a broad health level, many individuals may not definitively know their cholesterol or blood pressure status, and these could be checked with their primary health practitioner. Other health professionals such as optometrists and audiologists can identify visual or hearing problems.

A fundamental issue at play here, however, is that the proven risk factors for dementia are not widely known in the community or among health care professionals. The curricula for many healthcare disciplines are unlikely to have kept up with developments in this field. The other issue is where we started with the book – we are simply not used to thinking about our brains and, hence, about brain health. We are broadly aware of the importance of prevention as it relates to other precious organs, such as the heart, kidney, genito-urinary tract or lungs, even the state of our skin. There is a good chance you may know at least something about how to reduce your risk of many other conditions, such as diabetes. There have also been highly successful health promotion campaigns – for example the "Slip, Slop, Slap" messaging in Australia and New Zealand to protect against skin cancer. Perhaps what is needed now is some effective health promotion approaches to protect the brain, as well as a concerted effort to engage with health professionals to become more vigilant and provide advice. Our brains, and the miracle of conscious awareness they somehow create, are far too precious to ignore.

Bibliography

Bartlett L, Bindoff A, Doherty K, Kim S, Eccleston C, Kitsos A, Roccati E, Alty J, King AE, Vickers JC (2023) An online, public health framework supporting behaviour change to reduce dementia risk: interim results from the ISLAND study linking ageing and neurodegenerative health. *BMC Public Health* 23: 1886.

Bindoff AD, Summers MJ, Hill E, Alty J, Vickers JC (2021) Studying at university in later life slows cognitive decline: A long-term prospective study. *Alzheimers & Dementia Translational Research & Clinical Interventions* 7(1):e12207.

Brayne C, Ince PG, Keage HAD, McKeith IG, Matthews FE, Polvikoski T, Sulkava R (2010) Educatiopn, the brain and dementia: neuroprotection or compensation? *Brain* 133: 2210-2216.

Livingston G, Huntley J, Liu KY, Costafreda SG, Selbæk G, Alladi S, Ames D, Banerjee S, Burns A, Brayne C, Fox NC, Ferri CP, Gitlin LN, Howard R, Kales HC, Kivimäki M, Larson EB, Nakasujja N, Rockwood K, Samus Q, Shirai K, Singh-Manoux A, Schneider LS, Walsh S, Yao Y, Sommerlad A, Mukadam N. (2024) Dementia prevention, intervention, and care: 2024 report of the Lancet standing Commission. *Lancet* 404: 572-628.

Lenehan ME, Summers MJ, Saunders NL, Summers JJ, Ward DD, Ritchie K, Vickers JC (2016) Sending your grandparents to university increases cognitive reserve: The Tasmanian Healthy Brain Project. *Neuropsychology* 30: 525-531.

Stuart KE, King AE, Fernandez-Martos CM, Dittmann J, Summers MJ, Vickers JC (2017) Mid-life environmental enrichment increases synaptic density in CA1 in a mouse model of Aβ-associated pathology and positively influences synaptic and cognitive health in healthy ageing. *Journal of Comparative Neurology* 525:1797-1810.

Vickers JC, Kirkcaldie MTK, Phipps AJ, King AE (2016) Alterations in neurofilaments and the transformation of the cytoskeleton in axons may provide insight into the aberrant neuronal changes of Alzheimer's disease. *Brain Research Bulletin* 126: 324-333.

A Summary of Projection Space Theory

> 'The brain's most clever trick is not to process the world, but to place you in it."

Recent decades have seen outstanding advances in our understanding of the brain as thousands of neuroscientists elucidate how the various parts of the nervous system work and are impacted by disease and injury. However, we don't yet have a full picture about how neurons, assemblies of neurons and the massive interconnectedness of nerve cells give rise to subjective experience.

Projection Space Theory starts from an unusual angle: rather than asking *how does the brain produce consciousness*, it asks *what is the brain's job*? One possible answer is that the brain's fundamental evolutionary purpose is to place the organism into the external world – to make a creature that lives *out there* rather than feeling trapped inside a skull. The brain is not a passive receiver of sensory data but an active generator of a continuously refreshed representation of the organism in its environment. This representation is the "projection." Our subjective experience may be linked to this projection: we are essentially running a neurally-generated story about our existence in the world and our relationship to physical reality.."

The theory has several interlocking claims:

1. **The projection is the conscious experience** – not its product. The brain produces

an experience that is not *watched* by some inner observer. The projection itself *is* the mental state. There is no genie, no homunculus looking at the movie – the movie is what you are.

2. **The brain is information-making, not just information-processing.** Rather than taking in sensory data and producing a representation as output, the brain is continuously generating a model of the external world and using incoming sensory data to update and correct it. The percept is constructed outward, not assembled upward from sensory inputs.

3. **The projection space is extendable.** When you drive a car you feel the tyres on the road – the projection extends to include the vehicle. When you read a novel you inhabit the story. When you use a tool your sense of self extends to its tip. The projection is not fixed at the body boundary.

4. **The projection has multiple layers.** It encompasses physical space (where I am, what I can touch), social space (other minds, language, emotional filters), and temporal space (memory of the recent past, anticipation of the near future). All are continuously running simultaneously.

5. **When the projection fails, consciousness is disrupted.** Dementia, psychosis, and brain injury can be reframed as conditions that damage the projection machinery, explaining why the person loses their sense of being present in the world and disrupting a range of brain processes.

A lot of the business of our nervous systems is conducted at a subconscious level – the busy work of maintaining our homeostasis, mobility, coordination, reflexes all going on seamlessly most of the time. So, what is the point of having conscious states? A conscious state brings experience all together and allows our brain the capacity to exist in and interact with the external world, as well as to overtly monitor our internal mental and physical states. Our brain runs a multi-sensory and cohesive movie of its engagement with our environment, tools, other people, the body etc. The conscious part is the process of projection; there is no little person somewhere 'watching' the conscious account unfold.

The brain has multiple processes to help make sense of the external world – it is constantly making predictions about our perception and running checks against this in terms of actual output. Importantly, it takes several hundred milliseconds to construct the projection – it isn't instantaneous. This means that the projection is subtly delayed, but the brain is adept at producing apparently seamless perception so that we feel we are in the present and consciousness is 'instantaneous'.

Our conscious experience is a production of the brain, which works fine for most people most of the time. But this process can be subverted, as in the case of sensory illusions, or where we may be actively hallucinating conscious experiences that feel as real as reality. Dreams reveal how good our brains are at stringing together recent and past experiences in the service of a story.

Which model of brain organisation fits best with this new model where we are actively projecting into a range of spaces? There is a good fit with most existing models. For example, the Global Workspace Theory suggests a conscious experience that moves from moment to moment based on when information from sensory modalities 'ignites' a global workspace. The idea that reentrant/recurrent neuronal pathways is a consistent motif across the cortex, which is bound together by oscillating linked corticocortical pathways, may also suggest how the brain assembles a running account of the world. Projection Space Theory also aligns well with the concept that perceived reality is a 'controlled hallucination', a constructed state of the brain.

Projection Space Theory also helps to explain the ability of the brain to work through tools and technology, and how we might engage with other projecting brains through language and other forms of communication. This is how we feel embedded in stories, music, movies – how we are fully transported into other constructs of being. This has clearly provided survival benefits, especially as we engage with tools and technology to enhance our survival and living standards, as well as how we build culture and communities with shared purpose. However, this has a dangerous side, as more and more people become transported into their internet devices or virtual communities and worlds, at the price of brain capacities and resilience. Technological advances have developed to such an extent that we are at risk of becoming 'functionally cyborged', our cognitive capacities outsourced.

Our expertise at developing and using tools – the coming together of brain biology and technology – is both exciting and unnerving. We may need to consider putting guardrails in place in light of our brain's easy projection and immersion into technology, and the outsourcing of human brain work to the machines through advances such as AI. Given the importance of brain work and stimulation for its healthy function, over-reliance on technological solutions to intelligent work may lead to atrophy and reduced resilience when confronted by diseases and conditions such as dementia, stroke, and traumatic brain injury. We will need the right policies in place to ensure that the 21st century sees our brains flourishing and not being pushed back into their skull prisons. Conversely, stimulating and challenging our brain is likely to prove one of the most powerful ways to protect it against the ravages of conditions such as Alzheimer's disease.

www.ingramcontent.com/pod-product-compliance
Lightning Source LLC
LaVergne TN
LVHW022227200926
848481LV00006B/2156
9781764798211